WATER RESOURCES MONOGRAPH SERIES **8**

The Scientist and Engineer in Court

Michael D. Bradley

AMERICAN GEOPHYSICAL UNION
WASHINGTON, D.C.
1983

Water Resources Monograph Series

The Scientist and Engineer in Court

Michael D. Bradley

2000 Florida Avenue, N.W.
Washington, DC 20009

Library of Congress Cataloging in Publication Data

Bradley, Michael D., 1938-
"The scientist and engineer in court."

(Water resources monograph; 8)
Bibliography: p.
1. Evidence, Expert--United States. 2. Civil
procedure--United States. 3. Forensic engineering--
United States. 4. Forensic scientists--United States.
I. Title. II. Series.
KF8968.25.B7 1983 347.73'64 83-10582
ISBN 0-87590-309-6 (soft) 347.30764

10 9 8 7 6 5 4 3 2

Printed in the United States of America

CONTENTS

1 INTRODUCTION

The gavel goes down, the witness is called and sworn in: "Will you tell the truth, the whole truth, and nothing but the truth, so help you God?" Every court day scientists and engineers take this oath, yet few know the duties of an expert witness and fewer still know the procedures in a lawsuit. Unprepared for the courtroom, they watch a drama unfold without benefit of the plot (Ball, 1975). Courtrooms swirl with costume and ceremony. Jury, judge, and spectators assume roles: audience, director, trier of the facts. And the lawyers! Lawyers are the key dramatis personae. Protagonist and antagonist confront one another as lawyers argue their clients cases in the courtroom. They object endlessly. They cross-examine opposing witnesses mercilessly. They speak an opaque jargon of "laches," "remittitur," and "stare decisis." They speak Latin: "ab initio," "in pari causa," "lex loci actus." They speak old French: "estoppel," "fee simple," "voire dire." And they speak law-speak: "abuse of discretion," "clearly erroneous," "malice aforethought" (Mellinkoff, 1963). They call contrary versions of the same story true. They plead for understanding and compassion. They mix independent variables called "facts" and a dependent variable called "law" into an argumentative gruel for court consumption. Then it's over, and the judge delivers his or her opinion. One lawyer calls it the decision of the decade, a magnificent example of benign, reasoned law. The other darkly threatens to appeal all the way to the Supreme Court, if necessary, to relieve an onerous and oppressive injustice. All of this is fascinating theatre, but to the unprepared scientist or engineer, the drama has none of the charm and all of the clarity of a Japanese "Noh" play.

The technical professional can no longer afford the luxury of ignorance about the judicial process (Thomas, 1974). Court decisions increasingly settle disputes and lawsuits that have significant technical or scientific components. As courts resolve more and more cases involving science and technology, the need for analytic skills becomes more pressing (Mitchell, 1978; Rosen, 1977). In order to use analytic skills, the court must accept scientific and technical information, process it for its substantive worth, and form it into a legal decision. The traditional method of information gathering for use in the courtroom is by expert witness testimony. This means that the expert witness is a key member of the team of professionals who participate in the trial for either the plaintiff or the defendant. But the information available to the scientist or engineer is not always placed before the court in the most effective manner.

Often the basic problem is one of communication, that is, transferring knowledge to the court in a manner that increases its likelihood of acceptance and use. Communication must contain more than scientific or technical information. To be effective, communication must be a process of sharing information with an audience. In the context of a lawsuit, communication needs to be structured so that it can be used by the court to make a rational and effective decision. This means that the audience (the court) is more than a passive recipient of information; it must assimilate and use the information in its internal processes. Effective assimilation and use will occur when both the expert witness and the court understand the information.

Communication is not the only problem; intellectual and professional isolation is another. Few scientists have the opportunity to develop a sophisticated understanding of the legal system. The demands of a scientific career are heavy and time is rarely available for the concentrated study and observation such an understanding would require. On the other hand, the members of the legal profession face similar pressures in their careers. Scientific issues are arising more frequently, yet the total conjunction between law and science remains relatively small. While the overlap contains some of the most important issues facing modern society, the press of daily affairs often seems more urgent for both scientists and lawyers. The links among the scientific disciplines and the law are found anew in each lawsuit dealing with a scientific or technical issue; and the critical point at which law and science meet is usually the testimony of the expert witness. Often, it seems that little incentive exists for a careful analysis of the role of scientific information in law. Indeed, hydrologists are often amused when it is suggested that the law is an important consideration in the scientific practice of hydrology. Why would a hydrologist care about the law of water? More specifically, why would a hydrologist want to be part of the legal process or even volunteer to serve as an expert witness?

Although expert testimony may be vital to the outcome of the trial, little assistance has been available for the scientist or engineer who must serve as an expert witness. The role of the expert witness is rarely discussed among scientists and engineers: Only recently have sessions been devoted to the topic at professional meetings; few articles on the subject relate directly to the concerns of a potential expert witness (Lamb, 1977); and only one or two educational courses are available. Moreover, senior expert witnesses rarely have the time to share their experience with future candidates for the job.

Trial lawyers are in a similar position. A lawyer preparing for a trial is exceptionally busy. He or she must read reams of documents, research the law, develop a trial strategy, assure procedural consistency, prepare the documents, counter the opposing lawyer's strategy, and argue the case in court. Between these tasks, the lawyer may or may not have the time to prepare adequately an expert witness who is perhaps only one out of several involved in the lawsuit.

The end result has often been that the scientific expert witness learned his role through trial and error, by trying a certain behavior and seeing whether or not it was acceptable. Thus an expert might refuse to answer a proper question and be corrected by the judge, who in his role as procedural overseer would explain that proper questions are meant to be answered, and that refusing to answer could lead to a contempt of court citation. Lesson: Experts answer proper questions. Or an expert might provide a sound and balanced testimony which unwittingly presented the opposing lawyer with ammunition to discredit or disclaim the expert's facts. Lesson: Experts need to realize the strategic value of information. Or an expert might mistakenly assume that a lawyer or a layman is not competent to understand a theory or a scientific model and, upon cross-examination, find him or herself attempting to explain a theory that was considered obvious and true by the expert, but challenged as conjecture by the opposing laywer. Lesson: Expert testimony is open to cross-examination; therefore, an expert's information must survive rigorous and meticulous probings from the opposing lawyer. Such lessons often come at the expense of embarrassing and sometimes costly mistakes in the courtroom.

This book addresses the need of expert witnesses to have a single source of basic information about the use of scientific knowledge in court. The message is simple: Scientists possess knowledge that courts find necessary when deciding lawsuits, and their presence will be increasingly required in court. But to be effective, the scientific expert witness needs preparation to understand what happens in lawsuits, and to whom and why it happens. At a minimum, this requires that experts have at least a basic understanding of court structure and operation, an idea of the role of scientific information in the legal process, and some notion of the special problems of the expert witness. Although courts may seem arcane and complex to a scientist or engineer, actions are understandable when viewed within the context of the judicial process; although lawyers may seem unscientific and illogical, their words and actions also are decipherable.

This book should be useful to any scientist or engineer who expects to serve as an expert witness. But it is written with the hydrologist in mind, and with examples taken from the fields of hydrology and water resources. The science of water, hydrology, is one of the few disciplines that has a counterpart in jurisprudence, water law; and the increasing interplay between hydrology and water law well exemplifies the growing need for the expert witness who is trained in science, but has an understanding of law.

The development of water law is inherent in the "resource" nature of water. Water is scarce and must be allocated. Although the hydrologic cycle endlessly replenishes the earth's fresh water resources, there is rarely enough for all users at all times. Water is allocated, therefore, by two important methods: as a commodity, distributed by economic arrangements such as markets, prices, and user charges; or as property, allocated by law, rights, and regulations. In the United States, law has been the means historically chosen for allocation, and water law has ordered human use of water to allow and encourage desirable activities (which tend to be economic "goods" like making a profit or selling property rights) and to discourage or restrict undersirable activities (which tend to be economic "bads" such as pollution). The institutional framework of American water law is a complex interaction of property rights, economic forces and public regulation. Most of it arose out of the settlement of water disputes.

Nearly all water disputes develop for three reasons: a change in use, a change in the location of use, or a shortage created when users compete for limited supplies. Uses and perceived shortages take varied forms in different locations. An irrigator in the arid Southwest pumps ground water from an aquifer and spreads it on fields of crops. A power company builds dams to increase pressure so that turbines and generators can turn when water passes through the gate. A manufacturer or a mine operator diverts surface flows for use as a raw material or to wash and treat other materials. A city pumps large quantities through its mains for use in homes and commercial establishments. Any of these users may also store water for the future, either through a public project such as a multipurpose reservoir or through a pumping regime that "stores" underground resources until needed. But water is a dynamic and transitory resource and when it moves, or is made to move in order to be put to use, conflicts often arise among users.

Conflict over water use is by no means new. For instance, water allocations between the private and public interests in the Colorado River basin have a long and turbulent history of conflict between jurisdictions, uses, and population (Hundley, 1978). Federal and state governments develop and allocate water and water rights to constituent groups. But users are increasing while the river's supply remains constant. Agricultural interests need the same water that could be used for energy development; and urban populations search for water for expanding cities and suburbs (Weatherford and Jacoby, 1975; Cummings and McFarland, 1979). Furthermore, the Colorado River is an international resource, and conflicts with Mexico over the amount and quality of water delivered by treaty are increasing (Furnish and Ladman, 1975). Conflicts are also developing over the proper groundwater regimes at the state level, over the conjunctive development of both ground and surface water resources, and over the water rights of Indians (Clark, 1974; Bradley and DeCook, 1978; DuMars and Ingram, 1980). Even well-intentioned public policies that deal with rights, resources, and uses encourage conflict, especially in the new and errorprone fields of environmental assessment and long-range planning for energy development (Holland, 1975; Kilburn, 1976). Water has always been viewed as the key regional resource. Its allocation is economically important enough to guarantee ongoing conflict in use.

Many water use conflicts are resolved by legal rather than scientific means. Although water problems are analyzed by the methods of science and engineering, water allocations are legitimated by the methods of the law. Each method is important for socially acceptable solutions to water problems. A well-analyzed scientific solution that ignores legal rights or accepted ways of allocation or legally sophisticated solutions that ignore physical reality are both partial answers that confuse more than clarify. Optimum resource development will require more closely integrated legal and scientific solutions (Younger, 1977).

While an expert's ideas and testimony can be important for a resolution, no expert can be expected to offer insights into the entire problem and its many ramifications. Courts settle water lawsuits in trials, with or without the help of technical experts. The issue is whether scientific and technical information can help to settle water conflicts effectively. If better information is needed, then expert testimony will help courts reach decisions based upon scientific advice. Only the input, information, is within the expert's control; the decision is the court's responsibility. The expert, however, can provide the best available information and expect better decisions to result from a more informed court.

Furthermore, while hydrologists and engineers are finding many ways to participate in the legal and public policy processes, there are increasing opportunities for the use of technical expertise in newer and relatively non-traditional decision arenas. Many regulatory and administrative hearings or tribunals now seek the expert as a witness. Although these adjudications are informal and quasi-judicial, they deal with important problems of public policy, such as the determination of water quality standards or the issuance of a mine dewatering permit. Administrative and regulatory proceedings are judicial in form and tone but different from court trials. Rather than judges, administrative adjudications might have a hearing officer or a special master. The law at issue might involve applying rules and regulations instead of statutory interpretation. The conduct of the proceeding, however, would be similar to a court trial, including the process of fact-finding by expert testimony. Learning to give testimony in trials should also prepare experts to better understand administrative proceedings.

And experts are playing advisory roles to other parts of the policy and decision making system. Hydrologists and engineers pursuing careers in the public service frequently find themselves serving as independent experts and advisors to courts and decision tribunals. These experts play a more advisory and less adversarial role than do private experts hired by a litigant in a lawsuit. These "experts to the court" are valued by judges as a source of objective and neutral analysis. Moreover, hydrologists sometimes are assigned to positions as lay-judges in those state judicial systems that allow such appointments (Silberman, 1975). The traditional office of court-appointed watermaster has a modern counterpart in the special master or hearing officer who gathers information prior to major water lawsuits. The U.S. Supreme Court uses special masters. A famous example is Simon Rifkind, special master in the case of Arizona v. California. Mr. Rifkind developed the data and technical analysis needed by the U.S. Supreme Court to equitably allocate the waters of the lower Colorado River among the states of Arizona, California, and Nevada. Even the international legal system uses scientific and technical experts more than previously to gather information for advisory panels and international tribunals (Sandifer, 1975; White, 1965). While each opportunity differs in specifics, all are alike in a basic way. All require an understanding of how and when scientific information can be best used in legal proceedings. Learning to be an effective witness helps prepare experts for those likely opportunities that use expert witnesses.

2 LAWYERS AND COURTS

It is hard to imagine a modern public institution so fascinating and yet so misunderstood as a court of law. The American fascination with courts arises primarily from the apparent unique power and influence courts have in society. A judge has public authority found nowhere else, ranging from the power to compel personal appearances of almost anyone before the bench to the power to issue orders which drastically affect entire communities and states. Thus a court, in certain instances, can order local, state, or national officials before it to provide testimony or explain a course of conduct. And, in certain instances, it can order the restructuring of community school systems, the rehabilitation of state prison systems, or the diversion or reallocation of major water sources. Courts, however, seldom take any of these actions independently; they only respond to the issues brought before them by citizens, institutions, and governmental representatives. They generally act only to validate the legitimate actions of a citizen, official, or institution, or to require them to meet an already legally mandated responsibility. Courts essentially serve as the final arbiters of those societal conflicts which cannot be solved by other means.

If the power of the court is misunderstood, so is the legal process itself. Primarily, this results from misinterpretation of the legal process by both the journalistic and entertainment media, which is compounded by the general public's lack of real knowledge about the legal system. In the press, attention is paid most often to the dramatic--to decisions of the U.S. Supreme Court or to criminal trials or sensational personal injury lawsuits in local courts. Supreme Court decisions are solemnly announced by TV newscasters as if they were pronouncements from an oracle of the law; rarely do the media offer careful explanations or evaluations of the issues decided. In local trials, attention is focused upon litigants scurrying up or down the courthouse steps and color drawings of witnesses on the stand; few commentators attempt to explain the procedural aspects of the litigation or the significance of the trial itself. At best, media coverage of the legal process is incomplete and trivial; at worst, it is often misleading.

The entertainment media, too, have helped create a false impression of the judicial system. According to scriptwriters, most police officers and sheriffs are violent and bent on avoiding public responsibility, most lawyers are shysters interested only in fees and power, most judges are incompetent and bigoted, most juries are impressionable and swayed by the trivial, and most court decisions are cruel jokes against the normal American. The legal heroes range from a clever Petrocelli to a stoic Perry Mason, who act more as detectives than lawyers. Using sleight of hand and logical tricks, they manage to solve the case by ferreting out the truth and confronting the guilty party with lies and evasive testimony. The defense always wins and the trial, which takes about fifteen minutes of air time, apparently ends when the witness breaks down and tearfully confesses. Many Americans have false impressions of the judicial process or courtroom operation from the fanciful dramatization of the law.

The misleading or false impressions created by the media are compounded by the lack of education available to the general public. Few courses are available in any high school or college that explain the judicial process in any detail to a wide audience of students. Those available are often inaccessible to anyone but the student studying political science or other pre-law majors. And few students ever feel they have the time to study judicial systems in any detail.

It is no wonder then that the expert witness coming to the courtroom for the first time has little basic knowledge of the legal process, court structure, or the duties and responsibilities of those involved in the judicial system.

The Actors in the Court Room

Assume for a moment that you are to give expert testimony in court, perhaps in a water rights lawsuit. As you step into the courtroom you find a different and interesting world. Serious business is conducted here, that much you can sense by the bearing and demeanor of the people. But who are these people? Some are robed, some are uniformed and some wear street clothes. Some talk; some listen. Who is who in court and how can their roles be kept straight? The following section is a brief guide.

Officers of the Court

The officers of the court control and administer court proceedings. Their role is to guarantee a fair, just, and efficient trial.

The *judge* is the officer who is either elected or appointed to preside over the court and its proceedings. The judge is the senior legal officer in the court and determines the facts, rules upon points of law and the presentation and admissibility of the evidence. In all trials, the judge is clearly and unquestionably in charge of the proceedings, and, when there is no jury, is the ultimate trier of fact. Naturally, he or she wears a badge of rank, the black legal robe.

The *court clerk* is an elected or appointed officer of the court who, under the judge's instruction, administers the proceedings. In jury trials the clerk gives the entire panel of prospective *jurors* (sometimes called *veniremen*) an oath. By this oath, juror promises that, if called, he or she will answer truthfully any question concerning his or her qualifications to sit as a juror in the case. Any venireman who is disqualified by a valid reason can be excused from jury service. A person may be excused for many reasons, including having a physical handicap such as blindness or hearing deficiencies, having recent jury service or because of professional demands that cannot be avoided such as those of a physician or a schoolteacher. Then the court clerk draws names of the remaining veniremen and a jury is selected. After twelve or fewer jurors have been appointed by the judge, the court clerk will administer an oath to the jury. The jury is the trier of fact.

The bailiff is the court officer who keeps order in the courtroom, calls witnesses and takes charge of the jury when it is not in the courtroom, particularly when the jury has received the case and is deliberating a decision. It is the duty of the bailiff to see that no one talks with or attempts to influence the jurors in any manner.

The court reporter has the duty of recording all of the proceedings in the courtroom, including the testimony of witnesses, the objections made by the attorneys either to the evidence or to the rulings of the court and the listing and marking for identification of exhibits offered or introduced into evidence. In some states, such as Arizona, the clerk of the court has sole custodial charge of all of the evidence exhibits.

The lawyers are officers of the court whose duties include representing their respective clients and presenting the evidence on their behalf so that the jury or the judge may reach a just verdict or a fair decision (A.B.A., 1974). Since the lawyer's role is pivotal and often misunderstood, the section below is devoted entirely to his duties and intellectual orientation and outlook.

Additional participants in the courtroom drama are the parties to the lawsuit--the plaintiff and the defendant--and the witnesses. The plaintiff is the party who has filed the suit, alleging injury or damage committed against him by the defendant. The witnesses are those persons, other than parties, who have knowledge about the facts or issues in the case, and have been called by one of the parties to present their information to the court.

The Lawyers

Lawyers are by far the most important actors in the courtroom drama. Although the lawyer has many complex duties, the most important is representing his client in the lawsuit. This point is crucial. A lawyer is in court as an advocate to present his client's case to the court. Representation includes many professional responsibilities. The lawyer is in charge of the lawsuit from start to finish: conducting the earliest discussions of the problem at issue with the client, gathering the evidence to present in court, finding and briefing witnesses, helping determine the jury selection, conferring with the judge upon the range of questions to be decided in a trial, making the verbal presentation of the case before the court, defending the client's interests throughout the proceeding, and winning the client's lawsuit, although in any trial only one side "wins" while the other side "loses." In essence, one lawyer directs half of the lawsuit while another lawyer, an adversary, directs the other half.

A scientific or technical professional often misunderstands the advocacy role played by a lawyer in court. One reason for this is that the daily activities of most scientific or technical professionals are unlike courtroom advocacy. Lawsuits and trials are special decision-making processes with special rules and procedures. Courts resolve conflict; that is, people go to court because they want to ask a judge or a jury to settle a dispute (Carter, 1979). To resolve conflict the law has developed a trial procedure based upon confrontation and advocacy. In a trial, both sides to the lawsuit offer to the court their side of the controversy. Both sides obviously see the same set of circumstances differently. The lawyers for each side "advocate" their client's explanation to the judge and jury; that is, each lawyer presents his client's case in its most convincing light while trying to cast doubt upon or discredit the case of the opposing side. A trial uses advocacy because this method best develops the necessary information to resolve conflicts that defy settlement in any other way. The lawyer in court is the advocate who fights for the client's interests in a structured examination of the facts, the issues, and the law.

Trial lawyers do not search for universal truths or broad generalizations, instead they seek an outcome in their client's best interests. A lawsuit applies broad norms and rules to the conflicts brought by individuals to court. In contrast, most scientist direct their daily activities toward broadening an understanding of physical cause and effect (Revelle, 1975). Appreciating the differences in the roles of a scientific and technical professional and trial lawyer will help a potential expert witness understand part of what happens in court and why it happens that way. The role difference is a matter of both preparation and training.

The lawyer's training assures a certain professional approach to problems which is distinct from a scientist's approach. Lawyers are active participants in the process of working out solutions and accommodations to human or institutional conflicts on a case-by-case basis. As a matter of style, legal education differs from scientific education; the emphasis of each profession's practice is toward different goals and results. In some ways, the law and science are similar but at the level of daily conduct the differences are enough to cause an observer to pause. Two of the most important differences are analytical ability and conceptual orientation.

A scientist should make no mistake, lawyers have a great deal of analytic ability. The analysis is usually of different aspects of a situation or a problem than would attract a scientist's immediate attention, but the ability to analyze the remains an important professional requirement for legal practitioners. In fact, when professionally stated, the lawyer's analytic ability sounds familiar to that of a scientist; it sounds "scientific." To quote a law school bulletin:

> Analytic ability is that especially observable capacity of the . . . lawyer to distinguish A from B, to separate the relevant from irrelevant, to stay on the subject, to sort out a tangle into manageable sub-components, to keep separate the verbal symbol and its referent, to examine a problem from close range or long distance, to detect an answer smuggled into a premise or a supposed 'fact,' to frame the same problem in many different ways, to be ever skeptical as to what is 'fact,' to know the place of--and limitations upon--logic in decision-making, to be able to surround a problem perceiving it from many different angles at once. In acquiring these skills, the lawyer must come to understand the process of generalization and abstraction; he or she must learn to move easily back and forth between the abstract and the concrete, to synthesize and to particularize with equal ease and to recognize when the solution to a problem calls for more data and when it calls for a choice among competing values. (Bulletin, Stanford Law School, 1975-1976).

Scientists will recognize this as a general analytic process: neither linear nor quantitative, yet a searching and critical way to organize data, generate hypotheses and reach conclusions. It differs from scientific analysis in both form and content, yet it has an internal rigor and logic that makes it readily applicable to the problems that lawyers solve. Since this analytic process works well for lawyers it is important to recognize it as similar to but distinct from the more formal scientific methods that work well for scientific investigation.

Another fundamental difference between scientist and lawyer is conceptual orientation. To a large degree, lawyers are process-oriented. A lawyer values procedural safeguards and methods almost as ends in themselves. Many notions of justice, fairness, and equity are process-notions; that is, they incorporate a sense of equal process for all before the courts--equal opportunity to be heard, to present evidence, to confront witnesses and so on. Since the substance of law is applied by the trial process, procedural consistency is one of the key ingredients to obtain acceptable decisions from the courtroom. No one wants a legal system that changes the rules of the game in an arbitrary manner. An important notion of fairness in the law is that all parties will be treated equally and that the same procedures and rules that have developed over time because they are efficient and fair will continue to be applied for everyone. Process, and especially fair process, is an important end of the law.

In subtle contrast, the scientific process is much more outcome directed. Process matters in science, but is it not a strongly held value by itself. In scientific experimentation, processes change and within some boundaries of consistency and replicability, a new or changed process might bring about a more valued result or understanding. Measuring methods also change, hypothesis are modified, even whole approaches to a problem can be altered if an increased understanding or a more valid or generalized model results. The past is not written as directly into the processes of science and engineering as it is in the law.

Procedural consistency is important in law and law students are taught its value as part of their professional education. Law students read large volumes known as casebooks that contain edited opinions from appellate courts and use the material in what is known as the case-method or case-system of instruction. The practice is based on the theory that the best method for learning law is studying actual decisions and deriving from them an understanding of the court's reasoning process. By using cases students learn by inductive reasoning the general rules and principles most frequently applied in the past and most likely to be applied in similar future cases. By critically analyzing and comparing cases students determine the law and its relation to other cases (Brody, 1978).

Thus the law student develops a profound sense of past legal development, and an expectation that the development will apply in some modified way to future similar situations. In the advocacy process a lawyer applies generalized principles and beliefs, such as fairness and equity, to a specific instance or conflict. Litigation, the law's basic tool, aims to apply the law to individuals and the judicial process is most at home when it disposes of a conflict situation uniquely (Cowan, 1968). All court cases are different in that each situation or conflict occurs between individuals. The law applies general rules to find a solution to a specific problem instead of searching for a general explanation of all conflicts or events that arise.

Another way to express this idea is to stress that the law finds resolutions based upon precedent, fact, and feeling. The law discriminates on the basis of human feelings--for example, feelings of justice, the right resolution of a dispute, the best ordering of human affairs. Overall the law follows community values and develops rules and processes to lessen the conflicts among humans. In this effort, fact-finding and careful observation are subsidiary to the peaceful ordering of human relations. Equity, equality, reasonableness, good faith, due process, and the speedy resolution of conflict are valued legal principles. Sometimes the factual basis for a decision may be distorted or misconceived and the law perpetuates injustices; but on balance, the strength of the law lies in its consistency and ability to make decisions among the value choices facing litigants in court (Cowan, 1968).

All of this may sound confusing at first, but when seen in action during a lawsuit much of the inner logic of legal processes should become more clear. The distinctions found in the law that seem strange to scientists are usually rational when understood in their own frame of reference. Generally the legal process is a matter of dispute settlement instead of data gathering or rigorous quantitative analysis. But these are not opposite approaches to reality, instead they are complementary methods of thinking. Each can strengthen the other, and the increased use of scientific and technical expert witnesses is an indication of the importance of integrating legal and scientific philosophies. Expert witnesses provide factual information that leads the courts to more factual decisions. In order to perform well, an expert needs a basic familiarity with the court system and court operations.

The Structure of the Courts

The United States has a federal governmental system with one national and fifty state governments that all make, interpret, and enforce law. This creates both federal and state court systems--more specifically fifty-one court systems--that operate side by side. They are distinct because they are created under different national and state constitutions. It is important to understand where the courts originate and the differences among federal and state courts (Thompson, 1974).

Where do the Courts Originate?

Federal Courts. The Supreme Court of the United States is specifically created by Article III of the Constitution (United States Constitution, Art. III, Section 1). All other federal courts are created by Congress. Congress has created two kinds of lower federal courts: "constitutional" and "legislative" courts. Constitutional courts are created under the Authority of Article III, the judiciary article of the Constitution, and legislative courts are created under the Authority of Article I, the legislative article. A constitutional court gives its judges constitutional protection--they are appointed for life during "good behavior" and their salaries may not be reduced. The most important constitutional courts are the district courts and the courts of appeal. The district courts are trial courts in ninety-four districts throughout the country, with at least one in each state and the District of Columbia. Four specialized courts also have constitutional status: the Court of Claims, the Tax Court, the Customs Court, and the Court of Customs and Patent Appeals. A "legislative" court is established by Congress for a specific purpose and staffed by persons who have fixed terms of office and who can be removed for cause or can have their salaries reduced. Legislative courts include, among others, the Court of Military Appeals and the territorial courts. Figure 1 shows the structure of the federal court system.

State Courts. The fifty states have similar court systems although the size, quality, and complexity vary by state. Most of the public's legal problems are resolved in the state courts. Each state's constitution establishes a judicial system or provides for the legislature to structure one. The terminology differs among states but a pattern can be found. The lowest court is often a system of local justices of the peace. The balance of the system is composed of state trial courts, appellate courts, and a supreme court or its equivalent (Abraham, 1975).

Usually the lowest state court is the justice of the peace (J.P.) or magistrate. This official is elected or appointed to administer relatively minor matters at the local level. Often the J.P. performs marriages, attends to misdemeanors or minor civil complaints, and decides issues that involve less than $200. Few J.P.'s are lawyers.

The next highest level of ordinary state courts is the Municipal Court, often called City Court, Traffic Court, Night Court, or Police Court. Regardless of name the municipal court is the first court of record in the state's judicial hierarchy, that is, transcripts are made of all or most official court proceedings. By custom and law, jurisdiction is limited to $500 to $1,000 in civil cases, and to misdemeanors in criminal matters. Municipal courts provide litigants with fast, relatively inexpensive trials, and are usually staffed by judges with legal training.

The next court in line is the workhorse of the state judiciary--the County Court, sometimes called Common Pleas Court or District Court. These courts have general civil and criminal jurisdiction over three major types of cases and controversies: ordinary civil cases beyond the limits of the municipal court;

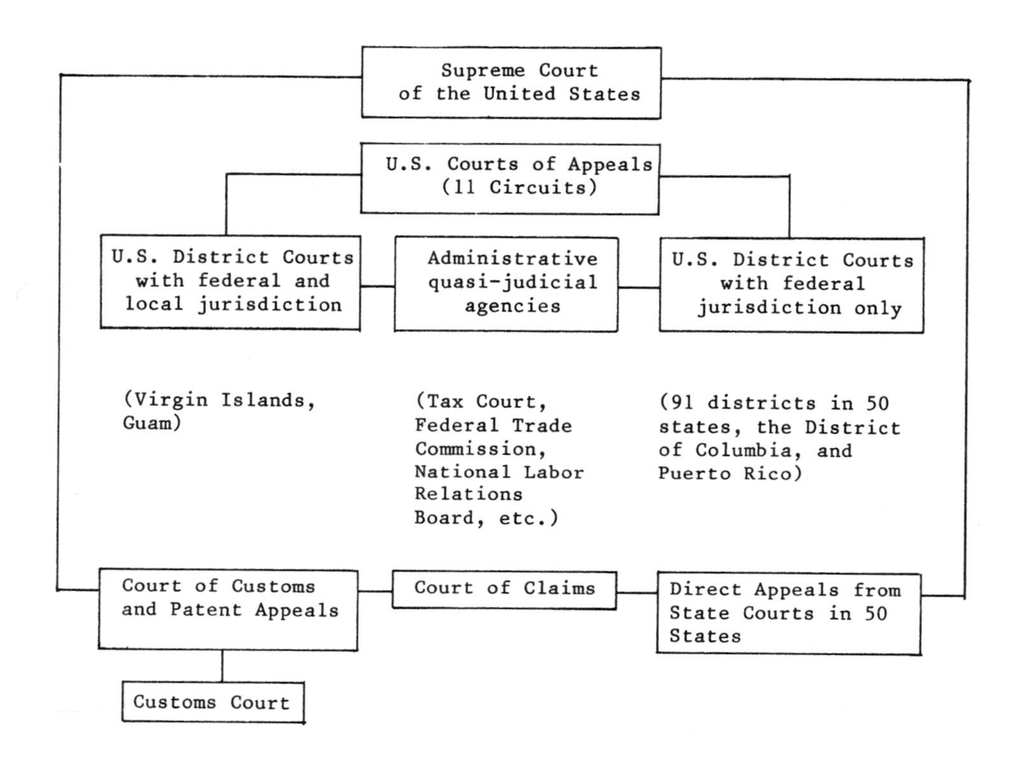

Source: American Bar Association, Law and the Courts, 1974.

FIGURE 1. FEDERAL COURT SYSTEM

criminal matters except for routine misdemeanors, and probate and inheritance. Jury trials are conducted at this level. As implied by its name, a County Court's geographical jurisdiction extends to the county line.

Above County Courts are intermediate courts of appeal, usually called the Appellate Division, the State Appellate Court, the State Superior Court, or the Intermediate Court of Appeals. These are not trial courts. They receive and judge appeals from the Municipal and County Courts, and usually their opinion is final. In unique instances, it is possible to appeal a matter beyond this court, but this is not automatic. Generally, the higher court has the final say in accepting an appeal from a lower court opinion.

At the top of most state judicial systems is a final court of appeals, sometimes called a Court of Appeals, a Supreme Judicial Court or a Supreme Court of Appeals. This court received appeals on major questions arising from lesser courts, often from the appeals courts. It will not accept cases that are merely concerned with factual issues; its main role is to find and interpret the law. An appeals court does not hear testimony or question witnesses or examine physical evidence because it assumes that the facts have been developed and decided in the trial. Instead, an appeals court decides upon the legal issues of an appeal. Figure 2 shows the structure of a typical state court system.

Court operation can be partly explained by structure, but it is also partly explained by jurisdiction. Jurisdiction is the authority vested in a court to hear and to decide certain types of cases. To understand court operation requires a familiarity with the basic jurisdictional divisions between the major court systems. The dual state-federal court structure complicates the task of describing what cases federal courts may hear and how cases from state courts may end up before the U.S. Supreme Court. These questions can be understood by examining the federal court system and its jurisdiction.

The Constitution of the United States defines federal court jurisdiction in the Judicial Article (Article III) and in the

State Supreme Court

(Court of final resort. Some states call it Court of Appeals, Supreme Judicial Courts, or Supreme Court of Appeals)

Intermediate Appellate Courts

(Only 23 of the 50 states have intermediate appellate courts, which are an intermediate appellate tribunal between the trial court and the court of last resort. A majority of cases are decided finally by these appellate courts.)

Superior Court

(Highest trial court with general jurisdiction. Some states call it Circuit Court, District Court, Court of Common Pleas, or Supreme Court.)

Probate Court

(Some states call it Surrogate Court or Orphan's Court. It is a special court which handles wills, estates, guardianship of minors and the mentally or physically incompetent.)

County Court

(These courts are sometimes called Common Pleas or District Courts. They have limited jurisdiction in both civil and criminal cases.)

Municipal Court

(In some cities, lesser cases are tried by municipal justices or magistrates.)

Domestic Relations Court
(Also can be called Family Court or Juvenile Court.)

Justice of the Peace
and
Police Magistrate

(The lowest court in the judicial hierarchy. These have limited jurisdiction in both civil and criminal cases.)

Source: American Bar Association, Law and the Courts, 1974.

FIGURE 2. A TYPICAL STATE COURT SYSTEM

Eleventh Amendment; by implication, all other matters are left to the state courts. Federal courts hear all cases "arising under the Constitution, the Laws of the United States, and treaties with foreign countries," called federal question cases; and cases involving citizens of different states or between citizens and foreign nationals, called diversity cases.

Some cases can be heard in either federal or state courts. For example, if citizens of different states sue one another for amounts over $10,000, they may sue in either federal or state court, or if a person robs a federally insured bank, he breaks both state and federal law and can be prosecuted in either state or federal courts. Lawyers are quite sophisticated at strategically placing their cases in the court that will give them better treatment. Prosecutors often file charges against someone who breaks both state and federal law in whichever court system is likely to give the toughest penalty (Wilson, 1980).

But some matters are exclusively under the jurisdiction of federal courts. For example, breaking a federal criminal law--but not a state one--will lead to a trial in federal district court. An appeal of the decision of a federal regulatory agency, such as the Environmental Protection Agency, will be heard before a federal court of appeals. In addition, a controversy between two states--as when California and Arizona sued each other over which state could use how much water from the Colorado River--would be heard only by the Supreme Court of the United States.

Complicated laws governing what matters reach the U.S. Supreme Court. Once many questions were open to Supreme Court appeal, but Congress passed laws giving the court control of its workload by selecting the kinds of cases it wants to hear. Now two routes (other than by starting there with original jurisdiction) are open to the Supreme Court: one is an appeal and the other is a writ of certiorari. An appeal concerns matters that involve clear constitutional issues, such as the constitutionality of conflicting federal and state law. Only about ten percent of the Supreme Court's cases arrive by appeal. The main route is by a writ of certiorari. This is an order issued by the Supreme Court to a lower court to send up the record of a case for review. The writ is issued when at least four of the nine justices vote for a review. The Supreme Court decides whether or not to grant writs of certiorari without divulging its reasons to the public (Wilson, 1980).

In special circumstances the U.S. Supreme Court may review the decisions of a state final court of appeals. The U.S. Supreme Court insists that "all remedies below" must have been exhausted before it will consider a request for review from a losing party in a State. But, if a federal question of substantial importance has been alleged and if that question has been properly raised from below, then a slim chance exists that the highest court in the land will accept the case for review (Abraham, 1975). Figure 3 shows how cases reach the U.S. Supreme Court.

Since the Supreme Court limits its cases and workload, the bulk of appeals from federal district courts (the trial courts) are decided by the U.S. Courts of Appeals. These appellate courts are between the Supreme Court and the district courts and are geographically organized by regions called Circuits. Figure 4 shows the regions of the circuits. Cases are heard by three to nine judges. Proceedings before the courts of appeal are conducted on the basis of the record in the trial court; that is, appeals are mainly questions of law decided on the basis of submitted documents (Abraham, 1978).

The federal system's workhorses are the trial courts, 94 district courts with about 400 judges. These courts were established by the Judiciary Act of 1789 to handle almost all civil and criminal cases arising from the federal law. It is here that the federal government begins and ends most of its prosecutions and suits. And it is here that the trial juries sit in the federal system; they do so in nearly half of all cases begun at this level. The district courts have original jurisdiction only; they are not appellate courts (Abraham, 1975).

1.) They may begin there (original jurisdiction.)

2.) They may arrive there on appeal.

   a.) From the highest state court, if the state court has declared a federal law or treaty to be unconstitutional or upheld a state law despite a claim that it violates a federal law or the U.S. Constitution.

   b.) From a federal court of appeals, if a state law or federal law has been found unconstitutional.

   c.) From a federal district court, if a federal law has been held unconstitutional and the United States was a party to the suit.

3.) They may arrive by a writ of certiorari.

   a.) From the highest state court where the case raises a "substantial federal question."

   b.) From the federal courts of appeals.

Source: Wilson, American Government, 1980.

FIGURE 3. HOW CASES GET TO THE U.S. SUPREME COURT

First Circuit:	Maine, Massachusetts, New Hampshire, Rhode Island, Puerto Rico
Second Circuit:	Connecticut, New York, Vermont
Third Circuit:	Delaware, New Jersey, Pennsylvania, Virgin Islands
Fourth Circuit:	Maryland, North Carolina, South Carolina, Virginia, West Virginia
Fifth Circuit:	Alabama, Florida, Georgia, Louisiana, Mississippi, Texas, Canal Zone
Sixth Circuit:	Kentucky, Michigan, Ohio, Tennessee
Seventh Circuit:	Illinois, Indiana, Wisconsin
Eighth Circuit:	Arkansas, Iowa, Minnesota, Missouri, Nebraska, North Dakota, South Dakota
Ninth Circuit:	Arizona, California, Idaho, Montana, Nevada, Oregon, Washington, Hawaii, Alaska, Guam
Tenth Circuit:	Colorado, Kansas, New Mexico, Utah, Oklahoma, Wyoming
Unnumbered:	Court of Appeals for District of Columbia

Source: Abraham, The Judicial Process, 1975.

FIGURE 4. THE REGIONAL JURISDICTION OF THE UNITED STATES (CIRCUIT) COURTS OF APPEAL

An understanding of the structure and jurisdiction of the various court systems provides important background to help the potential expert witness to understand the process of civil litigation. The various rules of civil procedure and their use are detailed in Chapter 3. At this point, however, a discussion of the factors involved in deciding to initiate a lawsuit is in order.

Going to Court

The first event leading to a civil trial is usually an alleged physical or financial injury to the potential plaintiff. In a complex, active society, both real and imagined injuries occur frequently, involving people, institutions, and machines. Most disputes that arise in everyday affairs, however, never reach the courtroom. Even when clearly definable legal rights and duties are involved, the injury or damage resulting from the dispute is minor and resolvable without legal action. On the other hand, some disputes result in injury or damage so severe as to result in legal action with no hesitation. These conflicts can be easily framed in legal language and effectively presented to a court. Falling between these extremes are the rest of the conflicts in daily life. While each is important to the individuals involved, the overall legal significance of the dispute may be questionable. Whether legally enforceable rights are involved must be determined before litigation is justified. The potential plaintiff will therefore seek a lawyer's help to determine the legal legitimacy of his complaint. If such a determination is made, the client may retain the lawyer and proceed with a lawsuit.

Whether a dispute is resolved by a lawsuit or not depends upon many factors, some within the control of the conflicting parties and some under the control of the court. A potential litigant has to decide to sue after an analysis of both kinds of factors, and a careful assessment of the strength of his legal position. Simple cases differ from complex cases in which the lawyer needs to conduct extensive research, preparation, and formal pleadings. The likelihood of the court ruling in a litigant's favor also must be assessed, as should the amount of damages likely to be recovered. These matters are all reasonably under the potential litigant's control and are essential to his decision to sue or not to sue.

In addition, several factors are structured outside of the litigant's direct control by the operation of the courts. Initiating a lawsuit does not automatically guarantee a trial because a lawsuit requires several threshold requirements of court procedure. For example, the conflict must fall within an area of court jurisdiction, and the court must have an available remedy that could resolve the situation. In other words, the court must have specific relief to provide the litigant. Also, a plaintiff must establish standing to sue; that is, the plaintiff must show that the injury directly affects him or her in a personal or economic way. Without standing to sue, a plaintiff cannot claim damages in court. These and other threshold requirements are used by the legal system to exclude some situations from trial for reasons that are partly a matter of court efficiency and partly legal custom or doctrine. So a willingness to sue is only part of the initial considerations involved in a successful lawsuit.

The Decision to Litigate

The first decision for a potential plaintiff is whether or not to litigate. While many direct and indirect factors help make the decision, a potential plaintiff must finally decide upon a course of action. The final decision may be rational and capable of advance calculation, or it may be emotional or irrational. Perhaps the most fundamental rational considerations are those that can be discussed in economic terms (McLauchlan, 1977).

Litigating means paying costs. A plaintiff must pay the costs of a lawyer, court fees, and expert witnesses as well as non-monetary costs such as losses in time, prestige, and psychological stress and strain. Further, the loser of the lawsuit may have to pay damages, a cost that is usually the largest of all. The balance of monetary and non-monetary costs is important in the decision to litigate.

The decision calculus cannot be expressed in numbers or a formula, but it can be written as an economic calculation. Assuming a cost minimizing strategy, a plaintiff will litigate rather than do nothing if the expected benefits of winning exceed the possible costs of losing (McLauchlan, 1977). The economic calculation for this decision is shown in Figure 5, Part I.

The most crucial element of the formula, the probability of winning (P), is partly objective and partly subjective. As P approaches 1.0 or certainty, the term PD-Cw increases and encourages a potential plaintiff to sue. But P has many uncertain elements, such as the legal precedents applicable to a situation, the governing substantive law, and the evidence or facts of the lawsuit. Other uncertainties are what judge or jury might decide after hearing the suit. So an objective evaluation of P is difficult and the trial lawyer needs to explain the risks of success or failure to a plaintiff not only at the beginning of a lawsuit but throughout the proceedings. P may change as the trial progresses and as opportunities for bargaining or out-of-court settlements arise or as time weakens both the memory of key witnesses and the sting of the injury suffered by the plaintiff.

According to the formula, winning is a matter of probability, not certainty. A litigant will seek minimum costs and maximum probability of winning. The total cost is easily fixed but the probability of winning or losing is largely a matter of conjecture. For example, if the probability of winning is approximately eighty percent, then the economically rational plaintiff will be encouraged to proceed with the lawsuit. If the probability of winning is only forty-five percent, then a plaintiff's expected benefits of winning will not exceed the costs of losing and a lawsuit should not be filed. Of course, these calculations are only part of the total strategic determination of a choice to sue or not. If the dispute involved a matter of important philosophical

FIGURE 5. THE DECISION TO LITIGATE

I. Generally, the plaintiff is a cost minimizer who will litigate rather than do nothing if the expected benefits of winning exceed the possible cost of losing.

$$PD - Cw > (1 - P)D + Ce$$

Where: P = probability of the plaintiff winning
D = damages
Cw = plaintiff's cost of winning
Ce = plaintiff's cost of losing

II. The defendant, once served with a complaint, can file an answer or do nothing.

$$PE > Cw$$

P = probability of the plaintiff winning
D = damages
Cw = costs of winning or not being held liable

III. A negotiated settlement is likely when both parties find themselves in a settlement range (X) that encourages settling out of court.

$$P_1D_1 - C_1 + S_1P_1 > X > P_2D_2 + C_2 + S_2P_2$$

P_1 = probability of the plaintiff winning

D_1 = damages sought

C_1 = cost of litigation for plaintiff

S_1 = cost of settling out of court for plaintiff

P_2 = probability of defendant losing

D_2 = damages payable

C_2 = cost of litigating for defendant

S_2 = cost of settling out of court for defendant

X = settlement range

or moral prinicīple or if the conflict needed settling for social reasons, then the calculation would have to be adjusted or extended to include these factors.

Since a lawsuit has two parties, the other litigant also has an economic calculation. The formula is Figure 5, part II. Once served with a complaint, this person must choose whether to hire a lawyer and file an answer or to do nothing. Many defendants do not file any answer or appear in court; instead, they let a default judgment be entered against them. This simplest of solutions means that if a defendant does not contest a suing plaintiff, he incurs none of the costs of losing, C_W. But the probability of losing becomes 1.0 or certainty as does the plaintiff's probability of winning. The defendant's certain loss will be equal to the damages being sought by the plaintiff, D.

If the damages sought by the plaintiff are greater than the costs of the defense, then an economically rational defendant will litigate. For example, if the defendant thought that he would pay less than $1,000 for lawyer's fees and litigation costs to present a successful defense, then he would be encouraged to litigate.

At this point, it is crucial to remember that these simple formulas have limited application. Both deal only with clearly identified money costs in the hypothetical example, and neither includes the consideration of other factors, such as the quality of legal advice, the emotional or psychological state of either party, or other intangibles. And the formulas only express what litigants might consider as an initial calculation regarding the dispute between them. Many disputes probably never get beyond the initial calculation by either the plaintiff or defendant because the possible gain may not justify further action against the other party.

If the decision is to continue the lawsuit, an economically rational litigant's calculations do not end at this point. When litigation seems wise, a bargaining or negotiation model more accurately describes whether a litigant should continue with the suit or seek an out-of-court settlement. This important point deserves repeating: a litigant in a lawsuit has several opportunities to make an important strategic choice--to continue or seek a compromise out-of-court settlement. This choice will become more clear after a discussion of civil procedures in the next Chapter. The economic formula for a negotiated settlement assumes a zone of self-interest (X in the formula) in which a litigant recognizes the strategic advantage of settling a lawsuit out of court. The formula is expressed in Figure 5, part III.

Several points about the formula need explanation. First, since each litigant is calculating his own costs and those of the opposing party, the calculations may or may not coincide. Each has different information and the lawyer for each party is the key element in reducing the guesswork involved in the formula's terms. Second, the probability estimates (P_1 and P_2) are likely to be independent of each other, even though the lawyers have nearly the same training and the same law to analyze. Clearly what is different is the skill or the experience of the lawyer doing the analysis. And, third, the litigation costs (C_1 and C_2) are fixed for each party. In the earlier formula used to decide whether or not to litigate, the party used two different cost factors, C_w and C_e. But, in the negotiations formula, each litigant has chosen a lawyer already so C_1 and C_2 are fixed costs (McLauchlan, 1977).

Thus, economically rational strategy involves a sequence of calculations. First, a determination is made whether or not a dispute should be litigated. If the decision is to litigate rather than not, then the parties file a lawsuit. But throughout the litigation strategic calculations are made as to the advisability of continuing the suit or negotiating an out-of-court settlement. Overall the economic calculation is a simple notion. It brings to the forefront an important result of most lawsuits. More disputes are settled by informal or out-of-court settlements than by full trials. Changing probabilities of winning or losing combined with strategic gains or losses in the trial often encourage litigants to settle their differences in more informal ways. The negotiated settlement is the most popular and effective dispute settling mechanism, and lawyers are important parts of negotiated settlements. For a potential expert witness, the role of the lawyer as a negotiator in out-of-court settlements is one of the most important legal realities to understand. Common knowledge notwithstanding, most disputes are settled by a lawyer's negotiations out of court.

The Lawyer as Negotiator

Lawyers would rather not take disputes to court. This statement sounds contrary to reality, but the logic underlying its premise is straightforward. Trials are expensive and uncertain. Most potential litigants want to avoid substantial lawyer's fees. Often they are well aware of the likelihood of losing the lawsuit and are quite agreeable to negotiating an out-of-court settlement. Even the most principled or determined plaintiff has to contend with a significant problem in modern law--crowded court calendars that result in lengthy trial delays. Delay can be an economic motivation because it encourages plaintiffs to discount expected winnings by a present value factor that reduces the value of an expected win. It may be better to accept a smaller settlement now than a less likely favorable judgment later. Defendants benefit from delay by enjoying the use of money that might be lost in litigation, earning interest in the present while decreasing the chance of a future damage adjustment against them (McClauchlan, 1979). Delay gives rise to important problems. Delay dilutes evidence: witnesses die, disappear, or forget. If a litigant has an important burden to prove certain points in the lawsuit, delay pressures him to settle. While the exact amount of delay is usually difficult to determine beforehand, most litigants include rough estimates of the effects of delay in their preliminary calculations.

Additionally, the experienced lawyer is often a major force for seeking a negotiated settlement. While some situations demand a lawsuit without hesitation, others are not so clear. Generally, a potential plaintiff seeks a lawyer's advice about a dispute before initiating legal action, and this will involve the defendant's lawyer in early discussions. Both lawyers perform a significant function--preventing, rather than encouraging, a lawsuit. This role is usually unrecognized by the press or the media. Most lawyers spend more time in their offices negotiating contracts or settlements of cases than trying cases in court (Brody, 1978). Either by careful drafting of legal documents or by strategic calculation and negotiation, most lawyers act as if one object is foremost in mind--preventing future litigation (Edwards and White, 1977). Although a direct estimate of the number of cases that would be litigated except for the foresight and care of lawyers is not possible, an indirect estimate is illustrated by the following figures: In 1973, 91.6 percent of the cases initiated in federal district courts were settled out of court prior to trial. Approximately 80 percent of all state court cases filed are settled prior to trial (Edwards and White, 1977). Much of the credit can be attributed to lawyers who controlled the situation from the time that action began and who encouraged an amicable or practical out-of-court settlement. (These estimates do not include disputes that are resolved among principals before going to a lawyer (Brody, 1978).

This lesson is singularly important for a potential expert witness. For one thing, it gives a realistic perspective to the expert who realizes that most disputes are resolved before a formal trial begins. Only an exceptional lawsuit will reach the trial stage. This encourages an expert to understand the many steps or procedures in civil litigation that give litigants a formal opportunity to settle their differences even while a trial is proceeding. (These are explained more carefully in the next chapter). And it allows an expert to structure his or her effort and fees on realistic expectations of out-of-court settlement instead of a long, drawn-out court battle.

This chapter has examined the structure and operation of courts from an external perspective. We have seen the structure of the court system, the actors involved in the trial process, and some of the factors involved in the decision to litigate. While the legal process holds surprises for the scientist or engineer, its logic is relatively straightforward. And, the legal process is full of strategic considerations, including the pervasive strategy of avoiding court altogether. In Chapter 3, the legal process will be examined from a more internal perspective that stresses the rules of civil court procedure.

3 THE PROCESS OF CIVIL LITIGATION

Courts operate by well-established rules of procedure. One of the important ways that courts regulate resolution of conflicts and problems is to apply a uniform set of rules to them. Different matters are structured by these rules into trials that are then "solved" by being heard and decided in court. The rules are formulated into a code, known as the Rules of Civil Procedure. Usually each state court system has such a code, as does the federal court system.

The goal of civil procedure is the just, prompt, and efficient settlement of disputes by the trial process. This means that procedures must perform serveral basic functions: notifying a defendant of a suit against him, informing each party of the claims and contentions of the other, determining the nature of the dispute and the issues between the parties, finding the facts in the case, deciding which principles of law govern the situation, applying the law to the facts to reach a judgment, and having the decisions of trial courts reviewed by higher courts of appeal (Cataldo et al., 1973). Usually, differences in court procedures are differences in the ways these basic functions are performed. American rules of procedure have shared roots in early English common law; consequently, they differ only in degree from early common law. A general examination of civil procedures has enough validity to be applicable nearly everywhere in the United States (Abraham, 1975). A summary table of the major steps in civil litigation is given in Figure 6.

As previously mentioned, a lawsuit results only after many calculations of success or failure. But if the lawyers are unable to

Defendant	Procedural Steps	Plaintiff
	Pleading Stage	
Motion to dismiss for failure to state a claim upon which relief can be granted.		Complaint
Answer	Motion for judgment on the proceedings	
	Pretrial Stage	
	Discovery	
	Interrogatories	
	Depositions	
	Motion for Summary Judgment	
	Pretrial Order	
	Trial Stage	
	Selection of Jury	
		Witness and cross-examination.
Motion for directed verdict		
Witness and cross-examination		
		Motion for directed verdict
	Post-Trial Stage	
	Concluding Arguments	
	Instructions to Jury	
	Jury Verdict	
	Motion for Judgment N.O.V.	
	Entry of Judgment	
	Motion for New Trial	

Source: McLauchlan, American Legal Processes, 1977, p. 78

FIGURE 6. PROCESS OF CIVIL LITIGATION

negotiate a settlement prior to trial, then a trial is likely to begin. A trial is guided by the rules of civil procedure.

Initial Rules: Starting a Lawsuit

A lawsuit is initiated when the plaintiff files a formal document known as a <u>complaint</u> in the court from which he seeks a remedy. (This document may also be called a declaration, a petition, or a statement of claim). The complaint is a statement of the basis and nature of the plaintiff's injury or damage and the relief or remedy sought. It typically contains allegations listed in numbered paragraphs which state the facts of the dispute and concludes with a "prayer for relief." This may be a simple demand for monetary damages or a request for equitable relief several pages long.

Once the complaint is filed, the defendant must be given notice of the action against him and brought within the court's jurisdiction. This is done by serving a <u>summons</u> on the defendant containing the name of the court, the name of the parties, and the name and address of the plaintiff's lawyer. A copy of the complaint is generally attached to the summons.

A summons is issued by the court clerk and served by the marshall, sheriff, constable, or other designated person. "Service" is normally ineffective unless the document is delivered personally, but alternatives have developed for serving an uncooperative or slippery defendant. These may include service to a defendant's dwelling or service upon any other person of suitable age and discretion residing there.

The summons serves to notify the defendant of the action against him and warns that unless the defendant takes certain actions to respond within a specified time (usually about 20 days), a judgment will be taken against him for the relief sought by the plaintiff. To avoid this default judgment, the defendant must answer the summons. This is done by making an <u>appearance</u> before the court. Appearance, in this sense, is a term of art and does not require the physical presence of the defendant or his lawyer before the judge. It requires only the submission of a formal notice to the court that the summons has been received and that the defendant is now represented by a lawyer. (If the defendant wishes to challenge the jurisdiction of the court over him, that challenge is usually raised at this point).

Once the plaintiff has revealed his complaint and his claims for relief the defendant has several options. He should immediately analyze the complaint and the plaintiff's case with a lawyer. Does the plaintiff have a sufficiently high probability of winning the damages requested? Does the complaint show a strong case? Has the plaintiff hired an extremely able lawyer with a reputation for winning civil suits of this kind? Will the plaintiff be willing to pursue this case through a lengthy trial and appeal, or will he or she accept an earlier out-of-court settlement? Can the defendant offer a strong defense or win a lengthy trial? The defendant may want to settle the action at this point. Perhaps the plaintiff will accept partial payment or binding promise in return for a release from the complaint. If so, then the lawyers can settle a figure (usually lower than the original damages) and enter into a written agreement sometimes called a stipulation. A copy is filed with the court clerk who records that the case is closed. Often lawyers spend more time and effort trying to negotiate a settlement between the parties than with the other aspects of the case.

But settlement is not the only action open to the defendant at this point. If the probability of winning is high enough, he can make motions that test the substance of the complaint and other important matters. The court's jurisdiction over the issue and over the defendant can be contested if this right has not been waived by an earlier appearance. In addition the formal sufficiency of the complaint itself can be attacked. If the complaint is ambiguous or vague, a motion might be made for a more definite statement. Or if the complaint has parts that are impertinent, immaterial, redundant, or scandalous, a motion might be made to strike them from the document. Or the defendant may question the legal sufficiency of the complaint. In early common law, this was called a demurrer. Modern codes now call the action a motion to dismiss "for failure to allege a cause of action" or "for failure to state a claim upon which relief can be granted." By making such a motion the defendant's lawyer is arguing that the court should dismiss the plaintiff's claim even if all of the assumed facts are true, because he is not legally entitled to a judgment in his favor. The motion is a timesaving device, and it makes an early legal challenge to the plaintiff's claim. It is also an opportunity for a strategic calculation, a reassessment of the probability of winning or losing in both the plaintiff's and defendant's calculus. Settlement is likely if the probability of winning changes or if the strategic assessment shows an increased chance of losing.

Once this challenge has been met, the defendant reveals his or her position with an answer which must be filed with the court within a statutorily specified period of time. An answer may take the form of a denial, an affirmative defense, a counterclaim, or some combination of these. Usually a denial will deny part of the plaintiff's allegations but admit others. A denial is important in civil litigation because it creates an issue which must be settled by the parties. If the plaintiff fails to prove a denied allegation, the lawsuit fails.

Even if the plaintiff proves all of the facts that would support a favorable judgment there may be special facts or circumstances which would deprive him of victory. If the defendant can allege facts in the answer that gives him advantage over the plaintiff, then he is making an affirmative defense. For example, suppose a plaintiff sues because a defendant failed to perform a promise contained in a written contract. Assume the defendant's answer admits that he or she entered into the contract and did not perform the promise. If no other facts are considered, the plaintiff should win the suit. But suppose the defendant has something to add--he or she was induced to sign the contract by the plaintiff's fraud. If this can be proved, the defendant may be released from an obligation to perform the promise. The affirmative defense is raised at the pleading stage of the trial and may change the probability of winning so much that the plaintiff will abandon the suit and negotiate an out-of-court settlement.

The third form of answer is a counterclaim. A counterclaim alleges facts that might have been asserted by the defendant had he sued the plaintiff. It may be based upon entirely different claims and may even include a demand for damages for more or less than the amount demanded in the plaintiff's complaint. A counterclaim is in effect a cross-suit. Usually an answer is the final pleading unless it contains a counterclaim; then the plaintiff is expected to respond to the counterclaim in the same way as the defendant answers a complaint. The response is called a reply.

The complaint, answer, and reply are known as the pleadings. The pleadings give an outline of the entire case. They must be carefully drawn because, although they can within certain parameters be amended, the substantive evidence admissible at trial will be limited generally to those issues raised in the pleadings.

Once the pleadings are closed, either party can move for a judgment on the pleadings. This is a formal device by which the court assesses the strength and validity of the claims of each side. It provides the parties with another opportunity to assess the strategic values of proceeding to trial or bargaining for settlement. Many cases filed in court are settled in the pleading stage.

Pretrial Procedures

Between the points at which the pleadings conclude and the trial begins, several events occur which will significantly affect the outcome of the case. Perhaps the most important of these is the pre-trial conference or hearing, now required by many judges in most complex civil suits. Typically, a judge has the lawyers appear before him to simplify and eliminate issues, admit facts, remedy defective pleadings by amendment, agree which documents are genuine, limit the number of expert witnesses, determine the scope of discovery, and decide if a master should be apponted to find information pertinent to the suit. Although the judge cannot order two parties to make an out-of-court settlement, he may at this point order them to engage in settlement negotiations and report back to him at a specified time. By forcing disclosure of facts, defining and limiting the issues, and encouraging settlement, the pre-trial conference serves to eliminate needless trials and streamline those cases which proceed to trial.

Although the conference may be informal in nature, the expert witness should not view it as a wasted or redundant procedure which he can or should avoid. If subpoenaed to appear at such a hearing, the expert witness, under penalty of contempt, must appear just as he must appear at a full trial.

Often considerable time is spent by the parties, witnesses, and lawyers establishing known and undisputed facts. This time can be saved by obtaining pretrial admissions. An admission is a statement made by a party or lawyer that a fact exists which helps the other side or that a point the other side is making is correct. Another important pretrial procedure is discovery. Discovery is the formal and informal exchange of information between sides in a lawsuit. Two methods of discovery are interrogatories and depositions. Interrogatories are written questions sent from one side in a lawsuit to another attempting to get written answers to factual questions. A deposition is the process of taking a witness's sworn testimony out of court, usually with the opposing lawyer given a chance to attend and participate. The expert witness, and any other witness who is called upon for a deposition, testifies as if he were in court and is subject to all the responsibilities and penalties of testifying in court (Cataldo, 1973). In general, discovery can extend to any matter relevant to the subject of the lawsuit whether or not it relates to the claims of the parties. The trial judge controls the nature and the extent of discovery proceedings and tries to prevent them from becoming "fishing expeditions."

The last pretrial procedure is the motion for a summary judgment. Again, as with the demurrer, this procedure weeds out relatively meritless lawsuits before trial; however, it differs significantly in form. A demurrer, or motion to dismiss for failure to allege a cause of action, strikes down suits which are insufficient as a matter of law, even if all of the facts are true. The motion for a summary judgment disposes of suits in which the facts cannot be proved. This motion can be used to attack an answer as well as a complaint and it can be the basis of another assessment whether to settle out of court or whether to proceed with the trial.

The Trial

Calendars and Jury Trials

After the pleadings are closed either side can file a notice of trial requesting the court clerk to put the suit on either the jury or the nonjury calendar. Normally a suit is tried by a jury only if the parties have a right to such a trial and if one of the parties takes the necessary steps to assert that right. Occasionally the judge may decide to use a jury. The right to jury trials is granted by constitutions and statutes. In federal courts, the Seventh Amendment of the United States Constitution guarantees the right to jury trials by providing that, "in suits at common law, where the value in controversy exceeds twenty dollars, the right of trial by jury shall be preserved . . ." (U.S. Constitution, Amendment VII). State constitutions have various provisions for jury trials.

When the time of trial arrives, lawyers will answer the calendar call and the suit will be assigned to a courtroom. In jury trials a jury will be empaneled. Jury selection typically consists of a voir dire examination of prospective jurors by the lawyers and the judge; that is, questions and answers intended to determine the qualifications of jurors. When a juror's answer indicates a bias, a prejudice, a financial interest in the outcome of the suit, a friendship with the opposing party, or any other disqualification, the lawyer may challenge for cause and state the reasons. Or the lawyer may make a limited number of peremptory challenges to excuse jurors without reason. Occasionally the judge will excuse an unqualified juror. The purpose of a voir dire examination is to empanel a fair and unbiased jury.

Once the jury is sworn, the plaintiff's lawyer usually makes an opening statement of the complaint. He or she will state the legal theory on which the plaintiff relies and sketch in the facts that the plaintiff intends to prove. Then the defendant's lawyer makes an opening statement, explaining the theory and the evidence for this side. Since the plaintiff usually has the burden of proof, he or she presents witnesses and evidence first. But a defendant who asserts an affirmative defense or a counter-claim may present first. Few trials follow exactly the same pattern and the judge has wide discretion to assure that the trial proceeds in the fairest and most efficient way.

The Nature of Evidence

Evidence is usually of two types: tangible objects called exhibits, such as photographs or rock samples, and oral testimony taken from the parties and other witnesses. Exhibits are used to develop important points in the testimony or to show physical evidence to the court. Testimony is how the court hears and accepts evidence from the observations and memories of witnesses.

The witness usually is called and questioned by one side of the lawsuit. These questions are called direct examination and their purpose is to place before the court and the jury facts supporting the case of the party who called the witness. Direct examination cannot include a leading question, that is, a question that shows a witness how to answer or that suggests a preferred answer.

Without this restriction trials would move faster but such tactics would substitute the thoughts of the questioning lawyer for those of a witness. The judge and the jury are entitled to know the thoughts of the witness.

Other limits are placed upon the kinds of evidence that can be considered and on the methods of presenting it. For example, when a lawyer asks a witness a question, the opposing lawyer may voice an objection to either the question, the answer, or both. An objection is a statement that the question of the other side in the suit is improper, unfair, or illegal. It asks the judge for a ruling on the point, that is, a statement that the question is wrong. The objection carries the request that the question or answer or both be stricken from the record and that the jury be instructed to disregard the improper point. Usually objections must be ruled on before the examination of a witness can continue. Most objections are based upon the legal principles governing the admissibility of evidence in a trial.

Rules of evidence. Rules for accepting or rejecting evidence are some of the most complex and difficult elements of the legal process. While a complete explanation is neither necessary nor possible, the expert witness, whose function is to present and interpret often complex evidence, must have some understanding of the most basic rules in order to be effective. A few of the most important rules of evidence are: the requirement of relevancy, the hearsay evidence rule, the best evidence rule, and the parol evidence rule.

One of the most frequent reasons for excluding evidence from consideration by the court is that it is irrelevant; that is, the evidence is not logically related to matters properly under dispute. Relevant evidence directly affects the issues in the pleadings and brings out the truth of the facts under dispute. Irrelevant evidence will not help to prove or disprove points that matter. Only relevant evidence is admitted or accepted.

Relevance is often the key factor in determining whether or not to admit circumstantial evidence in court. Circumstantial evidence is offered to prove indirectly a fact in question; for example, testimony that a person was seen walking in the rain is direct evidence that a person walked in the rain, but testimony that the person was seen indoors with wet shoes on a day when it was raining is circumstantial evidence that the person had walked in the rain. By way of another example, discovering a live dodo louse, a parasite specific to a host species, is circumstantial or indirect evidence that dodos are not extinct. Relevance is the determining factor and relevant circumstantial evidence can help prove many "facts" to the satisfaction of judges and juries even when direct evidence is scarce or when causal effects are uncertain.

Relevant evidence often is excluded from consideration by the hearsay evidence rule. Hearsay is second-hand evidence; the facts are not in the personal knowledge of the witness, but a repetition of what others said. It may be oral or written, but it is evidence that depends upon the believability of something or someone not available to the court. Although hearsay evidence could be excluded because its source is not under an oath of truth, a better reason is that the source cannot be cross-examined before the court and the jury. The hearsay rule has exceptions but usually it keeps direct testimony before the court and jury so that facts can be more easily ascertained and used in the trial.

The best evidence rule requires that the most reliable proof of a fact be produced for consideration by the court; for example, if a painting is available as evidence, then a photograph of the painting will not be the best evidence. If the contents of a document such as a contract, book, receipt, or will is to be proven, the best evidence rule requires the introduction of the primary source--the original document. If an original is difficult or impossible to acquire, exceptions can be made; but in situations where what is written is at issue, the best evidence rule generally requires an original document.

The parol evidence rule says that when two parties make a contract or a written agreement, the meaning of the written agreement cannot be changed by using prior oral agreements. In other words once the agreement has been written and accepted as final, complete, and accurate, other evidence of prior communication is not admissible for contradicting, amending, or changing the agreement. All prior understandings are assumed to have been incorporated into the written agreement. The reason for the parol evidence rule is that when parties take the time and trouble to agree in writing, they should be protected against a future distortion of their agreement by a dishonest claim.

Opinions and evidence. A fundamental principle of the law of evidence is that a witness should testify only to what is observed through the senses and not to conclusions that might be drawn from observations. For example, when discussing an oral agreement, a witness could not properly say that "We agreed..." because this would be drawing a conclusion. Instead the witness could testify that "I said..." and "he said...". No conclusion is then drawn by the witness and the determination and analysis of facts can be left to the trier of facts, the judge or jury.

But not all evidence is a matter of direct experience or observation; an opinion can be an important source of information for the court. An ordinary witness cannot generally give opinions as testimony. Opinion evidence comes from experts who have special knowledge and experience. The testimony of expert witnesses will be detailed in Chapter 4.

Cross Examination

After a witness has presented testimony and been questioned on direct examination, the opposing lawyer cross-examines. Cross-examination is for the purpose of limiting, explaining, or refuting statements made on direct examination so that incorrect interpretations or undue weight are not given to them. When portrayed on television cross-examination is often dramatic: Dishonest witnesses are ensnared in a tangled and revealing web of lies and Perry Mason guides a naive and trusting jury to the truth. Real life is rarely so dramatic. In actual cross-examination the methods used and the result achieved are usually more sophisticated and significant than what an outsider would expect. Parts of the truth can be revealed by cross-examination that may have been ignored, forgotten, or strategically avoided by direct examination. Even honest witnesses may admit a lack of direct observation of facts presumed true, or a deficiency in hearing or eyesight that would modify the results, or an inaccurate expression in earlier testimony. It is important to understand that cross-examination is one of the major truth-finding mechanisms in the legal process and that its use is one of the most important strategic tools of the trial lawyer. Cross-examination proceeds according to the rules of evidence and allows each side to examine the same situation of fact from its own viewpoint. A more complete truth is likely therefore to emerge than if only one side presents evidence without examination.

Motions at the Close of the Plaintiff's Case

After the plaintiff submits all of his evidence and after the completion of the direct questioning, cross-examination, and redirect examination of each witness, the plaintiff usually rests. This is another crucial moment in most trials, for now the defense has a full measure of the case against it and another opportunity to figure the costs of proceeding. If the calculation is positive and the defense decides to proceed, then the defendant can make some strategic motions: a motion for a directed verdict, or a motion for voluntary nonsuit.

Motion for Directed Verdict

At the close of the plaintiff's case the defendant usually calls on the court to decide the sufficiency of the plaintiff's evidence by making a motion for a directed verdict. In contrast with the earlier demurrer which is granted if the plaintiff fails to allege a cause of action, the defendant's motion for a directed verdict claims that the plaintiff has failed to prove his case. For this motion to be granted the judge must agree that the evidence, even if true, fails to establish an adequate basis for recovery. If a fair and unbiased jury would not find for the plaintiff, then the motion should be granted. If granted, the plaintiff loses the action <u>and cannot start another</u>; if denied, the defendant proceeds to offer his evidence.

Motion for Voluntary Nonsuit

Sometimes the plaintiff, having presented all of his evidence, realizes that the evidence is insufficient and the case is weak. Given a second chance, the plaintiff could fill the critical gaps in the evidence but needs to avoid having judgment taken with no right to start another action. He can move for a <u>voluntary nonsuit</u>. If granted, the plaintiff can begin another action after paying the costs of the original proceeding. A voluntary nonsuit can be pleaded at any time between the trial's start and the time the judge either submits the case to a jury or directs a verdict against the plaintiff.

If the defendant makes no motion for a directed verdict or if the motion is made and denied, then he may introduce evidence for the defense or the defendant may rely on the weakness of the plaintiff's case and expect a voluntary nonsuit or a favorable jury judgment. Should the defendant choose not to introduce evidence, the plaintiff can move for a directed verdict on the theory that the present evidence would favorably convince a jury. Since the defendant had a fair opportunity to present a defense, the plaintiff may now move for a judgment.

The Defendant's Evidence

Usually the defendant does introduce evidence, and this process may begin with an opening statement, if one was not made earlier.

The procedure is the same as for the plaintiff's evidence except that the defendant's lawyer will directly examine witnesses and the plaintiff's lawyer will cross-examine. After the defendant's last witness the plaintiff may move for a directed verdict just as before. This time the judge must consider the evidence on both sides in deciding whether a reasonable jury would find for the defendant. If the motion is denied, then the plaintiff may offer evidence in rebuttal. The defendant introduces evidence to meet any new matter until both parties rest. Finally both parties are entitled to make another motion for a directed verdict. If denied, the trial continues.

Closing Arguments or Summations

The next step is for the lawyers to make closing arguments to the jury, a process also known as summing up. In most cases the plaintiff's lawyer addresses the jury, the defendant's lawyer does the same, and the plaintiff's lawyer speaks in final rebuttal. In summing up, each lawyer recapitulates the claims, summarizes the evidence supporting his arguments, states the legal principles supporting his case and comments upon the evidence--its credibility, weight, and believability.

Instructions to the Jury

After the closing arguments, the judge charges the jury; that is, he instructs them how to proceed in their deliberations. The judge reviews the closing arguments, points out the most important issues of fact, summarizes the testimony and the evidence and explains how the jury should weigh it. Many of these matters are routine but other matters may be unique to the trial. These are usually offered to the judge by requests to charge submitted in writing by the lawyers and supported by appropriate citation to legal authority. Oral requests relating to the charge may also be given. Each lawyer may object to the court's instructions to the jury.

An important matter for jury instruction is the burden of proof. The plaintiff has the burden of proof for facts that establish the right to recover and the defendant has the burden for affirmative defenses and counterclaims. The burden of proof must be by a preponderance of the evidence. The preponderance is not a majority of witnesses or any special kind of evidence. It requires that the jury find the evidence for the burden of proof more credible and convincing than the contrary evidence.

The Verdict

Once charged, the jury goes to the jury room to deliberate and to reach a verdict. Some jurisdictions no longer require a unanimous verdict. If a jury cannot reach a verdict there is a hung jury. A hung jury requires a new trial and a new jury. Usually juries return verdicts according to the facts presented in the trial and the judge's instructions.

Motions After the Verdict

After a verdict there is little more to the lawsuit than the entry of judgment. Sometimes the losing party may make either a motion for a judgment notwithstanding the verdict, called a judgment N.O.V., or a motion for a new trial. A judgment N.O.V. requires the court to review the evidence and make the same determination as if deciding a motion for a directed verdict. If the judge thinks that a fair and unbiased jury could have decided only for the party making the motion, the judgment could be granted notwithstanding the verdict. Otherwise it should be denied. If granted, one might well ask why a motion for a directed verdict was not granted in the first place rather than requiring an entire trial and a jury verdict. There are many reasons. For instance, the judge might be more certain of the legal authority to act now that the trial has finished and all the evidence and summaries have been heard. A more likely reason is that less harm is done by granting a judgment N.O.V. erroneously than by granting a motion for a directed verdict erroneously. If a trial judge errs in granting a motion for a directed verdict, an appeal requires a new trial with all of the expense of starting over again. If a motion for judgment N.O.V. is granted erroneously, all that is required to right the error is a direction to reinstate the original verdict.

If the motion for judgment N.O.V. is denied, the unsuccessful party may move for a new trial. This motion might be granted for several reasons: newly discovered evidence, excessive damages, surprise in the trial against which the losing party could not have guarded, error by the court during the trial, a verdict against the weight of the evidence, or others. The motion for a new trial is given when it appears that a reasonable jury might have found a verdict counter to the one set aside. These procedures give judges broad power over the verdict of juries and prevent unreasonable or abusive verdicts from juries that might be biased or influenced by outside forces, such as a hostile press or media. Few judges exercise these powers to the fullest. Usually the jury system works well in the instances where it is used.

Many trials are held before a judge alone, eliminating the possible problem of using a relatively unpredictable jury as the trier of fact. Generally the same rules and procedures apply to a trial before a judge alone as to a jury trial. The few differences in these trials are a matter of procedure relating to the jury selection and use. A trial before a judge alone requires no jury selection, briefer opening statements, and more frequent and open conferences between lawyer and judge. In nonjury trials the rules of evidence apply but a judge is less likely than a jury to be influenced or confused by dramatic or improper evidence. Nonjury trials need no instructions at the end of the presentations; instead, a judge issues specific findings of fact, specific conclusions of law, and specific judgments. Without a jury, a motion for a directed verdict or for a judgment N.O.V. is useless, but a motion can be made for a voluntary nonsuit or for a new trial.

Judgment

If the verdict of the jury or the findings of the judge are not set aside and if no new trial is granted, then the court directs the entry of a final judgment for the successful party. Where the amount is definite, the judgment will probably be for the amount claimed including interest. Indefinite claims usually are for less than the original sum.

Appeals

In addition to trial courts most judicial systems have a network of appeals courts to which a party may appeal when he thinks a lower court has made a mistake. Appeals courts not only assure that individual cases are properly decided but also keep the law uniform by settling differences of opinion over legal interpretations by the lower courts. For an appellate court to review the actions of a trial court an _appeal_ must be filed by the complaining party. An appeal asks a higher court to review the decisions of a lower court in order to correct mistakes or injustices. The person who appeals a case from a lower court is known as an _appellant_, and the party in a case against whom an appeal is taken is known as the _appellee_. Usually the appellant is the person against whom the judgment was rendered in the lower court, but successful parties appeal occasionally, especially if they think the amount of the lower court's award too low.

To appeal the appellant must file notice with the trial court clerk and serve notice on the appellee. The appellant must also post bonds to cover costs if the appeal fails and to secure the appellee for the amount of the judgment, in case the appellant's assets are dissipated while the appeal is pending. This also bars the appellant from enforcing the judgment until after the appeal.

Next the appellant prepares a record on appeal. Since appellate courts only consider matters appearing in the record, the appellant must take care to furnish an adequate record. The appellant's lawyer files a brief with the court clerk and several copies with the opposing lawyer, who responds in the same manner. The briefs inform the appellate court and the opposing lawyer of each party's contentions and of the legal authorities supporting each side's argument. If the appeal is from a judgment without a trial, all that is required is a statement of legal propositions and the legal support for the propositions. If the appeal is from a judgment after a trial, it is also necessary to relate portions of the transcript to each proposition being contended. Appellant and appellee may submit the case to the court based solely on their written briefs or they may request oral argument. In oral argument, an attorney for each side briefly synopsizes or "argues" the substantial issues of the appeal and answers questions from the bench. No witnesses are heard. Appeals courts rely upon the record of the trial in lower court for an understanding of the facts in the trial. In general, appeals judges do not consider facts not already in the record.

Generally an appellate court considers questions of law and will not review a trial court's rulings on questions of fact. Trial courts are assumed to have adequate access to the facts of a dispute because the entire adversary process of evidence, testimony, and cross-examination is a procedure for generating and revealing the truth as seen by both parties. An appeals court hesitates to second-guess the facts at issue and contested in a trial court. Instead appeals courts examine the lower court trial for reversible errors in the law only, and are generally reluctant to overturn a lower court decision.

In a single trial many rulings are made that may affect the outcome. Judges are the most likely source of error since rulings from the bench are made throughout the proceedings and each ruling may unjustly advantage one side or the other in a lawsuit. Before the trial, a judge may rule on a motion to dismiss for lack of jurisdiction or for failure to allege a cause of action. During the trial rulings may be made on jurors, admissibility of evidence, conduct of the trial, instructions to the jury, directed verdicts, judgment N.O.V., a new trial and other matters. Each ruling is a potential source of error.

When found, errors in law may or may not be reversible. If the error is harmless because it was not prejudicial or because it was neutralized or corrected, then a reversible error did not occur. A reversible error is one that might have prejudiced the decision against the appellant.

Once an appeal is heard the appellate court may affirm, reverse or modify the judgment of the trial court or it may grant a new trial. Appeals decisions require a majority vote from the judges who hear the case. Once decided, the appeals court sends the case back to the trial court for whatever action is necessary. Appeals court opinions are published periodically in bound volumes. Along with statutes, these opinions are a major source of the law (Cataldo et al., 1973).

In summary, the rules of civil procedure guide and direct a lawsuit when it is being tried in court. Many of the rules have origins in the development of early common law. Other rules are more recent and try to deal with changing types of evidence and trial procedure. All of the rules have the ultimate objective of assuring the prompt, just and fair settlement of lawsuits brought before the courts. An expert witness needs a basic understanding of the rules of civil procedure for two reasons. First, an expert's effectiveness will in no small measure come from an understanding of the process of civil litigation. And, second, an expert needs to assist the lawyer in the presentation of factual evidence needed by the jury and the judge to make a proper and enlightened decision. How an expert performs these roles is the topic of the next chapter.

4 EXPERTS AND EVIDENCE

As courts increasingly face problems requiring scientific knowledge or specialized experience, the need for expert witnesses grows. Experts have the knowledge, training, and experience to help a court understand questions which, without assistance, an inexperienced person is unlikely to appreciate fully or correctly. The value of the expert lies not in his ability to place facts before the court, but in his ability to provide an analysis of those facts and inform the court of the scientific and technical consequences of the issues at hand.

But the expert witness must testify in court like other ordinary witnesses; and rather strict rules have developed regarding the testimony of witnesses. Exceptions to some of these rules are made to facilitate expert testimony. Usually these exceptions can be taken advantage of only if the subject is one where the opinion of a skilled and experienced expert has greater validity than the opinion of an ordinary juryman, and if the witness is properly qualified as an expert (Rosenthal, 1961; Patch, 1978).

The Historical Development of Witnesses

The role of modern expert witnesses evolved from the common law courts in medieval England. Any distinctions made between an expert and an ordinary witness are the result of legal history, the development of science and the scientific professions, and the important role courts have in settling social conflicts. The use of witnesses as sources of factual evidence in a trial did not really develop a common law until the use of a jury as a fact finder was firmly established. Originally, a "trial" operated as a submission to a mechanical process of proof, and the defendant who survived was deemed innocent of his alleged offense. Less violent forms of justice required the defendant to produce a specified number of people who would back the defendant's denial of wrongdoing. Witnesses, when used at all, were important only for their willingness to take an oath that the defendant was innocent, and not for any knowledge they might have.

Witnesses and juries were a slow development in common law. Most of the framework for royal courts and administration was built between 1066 and 1087 by the Norman King, William the Conqueror, who developed an advisory body known as the royal council and the first demographic and economic census known as the Domesday Book. Data was gathered for the census by a Domesday inquest, a process which used juries of local residents to supply information. Later, during the reign of King Henry I (1100-1135), royal judicial administration grew significantly. Local justices in each shire assisted the sheriffs with more routine disputes, while important cases were judged by the King and his great men at the royal court. Royal justices were sent to various parts of England to hear pleas in the King's name, enlarging the scope and effectiveness of royal justice. The judicial tours began in an erratic manner, but eventually became a comprehensive, regular part of judicial administration.

Juries did not become a regular part of the developing legal system until the reign of King Henry II (1154-1189). And, initially, juries, formed by selected men from each town, served only to supply the names of local criminals or wrongdoers; the accused were then forced to submit to ordeals and were suitably punished if they failed to pass (Hollister, 1976).

Early trials were more mechanical than their modern counterparts; that is, the process of proof depended less upon objective and empirical information and more upon the outcome of the ordeals or of the trial itself. There were four major types of proof: witnesses, compurgation, battle, and ordeal. A trial decided with proof by Witnesses was relatively simple, an accused produced witnesses to swear that they believed his story. The oath was important, but not its truth. In a trial by Compurgation, a defendant denied the charge against him and produced a certain number of persons or "compurgators" to back the denial with their oaths; if he had a sufficient number of compurgators, he would win his case. The sufficient number varied with social standing and the severity of the accusation, for instance, a highly ranked individual would need fewer than a peasant, and a simple thief fewer than one accused of murder. Trials by witnesses and compurgation were relatively straightforward, and inexpensive. Sheriffs and justices could decide cases soon after accusations were made and within the shire or tourship where the dispute originated, although sometimes these issues were settled in the royal court.

Two other methods of deciding trials were more elaborate since both required the intervention of Divine Providence to help decide which party was right. The trial by Battle was settled by combat. The accused and the person accusing could meet on the field of honor or they could hire champions to battle in their place. Victory was the reward of the most righteous and the loser was clearly guilty. And the final form of proof was trial by Ordeal. Since God was on the side of the innocent, then heavenly intervention would provide some sort of sign or miracle to help determine guilt or innocence. The court's role was to choose an appropriate ordeal, carefully follow established procedures, and observe the result. The famous legal commentator Lord Blackstone described this form of trial:

> The most ancient species of trial was that by ordeal . . . This was of two sorts, either fire-ordeal, or water-ordeal; the former being confined to persons of higher rank, the latter to the common people. Both of these might be performed by a deputy, but principal was to answer for the success of the trial; the deputy only ventured some corporeal pain, for hire, or perhaps for friendship. Fire-ordeal was performed either by taking up, in the hand, a piece of red-hot iron,

> of one, two, or three pounds weight; or else by walking barefoot, and blindfolded, over nine red-hot ploughshares, laid lengthwise at unequal distances; and if the party escaped being hurt, he was adjudged innocent; but if it happened otherwise, as without collusion it usually did, he was then condemned as guilty . . . Water-ordeal was performed either by plunging the bare arm up to the elbow in boiling water and escaping unhurt thereby, or by casting the person suspect into a river or pond of cold water. If he floated therein without any action of swimming, it was deemed as evidence of his guilt; but if he sank, he was acquitted." (Black's Law Dictionary, 1968).

Gradually, however, the jury trial replaced the older forms and the use of information developed in different ways. Legal decisions resulted from the reasoning process of a group of rational men contemplating the available information rather than a mere submission to a mechanical test. Early juries were likely to be groups of neighbors already acquainted with the facts under dispute or capable of easily discovering them. They judged the character of the witnesses as much as the facts, since there was no settled practice for obtaining information by means of witnesses. It was not until 1562 that Parliament compelled witnesses to attend and testify in the common law courts.

As the jury began to function more widely in the judicial system, occasions arose where the tribunal needed knowledge or information in order to decide issues reasonably. The required specialized knowledge could be obtained two ways: one was to empanel a jury of persons specifically qualified to pass judgment on a particular case, this was a jury of experts; the second method was to summon to court skilled persons to inform the jurors about those matters beyond its knowledge, this was an expert witnesses. The court could instruct the jury to use the knowledge or to judge its usefulness in their own findings.

In all likelihood the need for specialized knowledge was first met by means of the special jury since ordinary juries were often acquainted only with local matters and problems. During the fourteenth century special juries decided questions dealing with trade and crafts; for example, juries were empaneled to decide whether the meshes of fishing nets were smaller than required by the trade ordinance, whether hides were properly tanned, whether tapestry was false, whether hats and caps were improper, whether wine was false, whether putrid victuals had been sold and whether a surgeon was guilty of malpractice. One special jury of married women (matrons) was empaneled to determine if a convicted prisoner was pregnant. (Rosenthal, 1961).

Early courts also summoned skilled persons to help them understand specialized problems. Documented examples deal with physicians testifying on medical matters. In a 1353 trial the court ordered the sheriff to summon surgeons from London to help decide if a wound was mayhem. Experts summoned under these conditions were probably asked to inform the court rather than the jury and the court would then instruct the jury. In effect, these witnesses were regarded as expert assistants to the court since the development of juries that decided issues from their own knowledge and that took testimony from witnesses did not evolve until later in the sixteenth century.

Thus the expert witness prototype was forming but there was no equivalent to the modern expert witness. The expert witness only became important when the proof of facts, rather than the jury's personal knowledge, became the accepted standard for a decision. These conditions gradually evolved during the sixteenth and seventeenth centuries in cases that usually involved the expert opinions of physicians.

By 1678 cases were heard with witnesses testifying on one side or the other and identified as the hired experts of a particular litigant. In the eighteenth century the expert's identification with one litigant became firmly settled with the decision in an important lawsuit, Folkes v. Chadd (Thomas, 1974). This case not only helped establish the admissibility of expert testimony it also decided an issue of water resources development. The question was the cause of a harbor filling. The plaintiff hired a respected hydraulic engineer to testify about the resulting injury to the plaintiff. Although the defendant objected to this evidence and argued that it was a matter of mere opinion, the court accepted the expert's testimony saying that, ". . . in matters of science the reasonings of men of science can only be answered by men of science." (3 Doug. 157, 99 Eng. Rep. 589 [K.B. 1783]). From this beginning the need for skilled assistance was met by the expert witness hired by one of the contending parties. Thus the evolution of modern expert witnesses had three stages: special juries, experts called to aid the court and experts hired by the contesting parties (Rosenthal, 1961). The last is the best known to modern courts.

Witnesses and Evidence

As use of witnesses developed, a more substantial law of evidence also emerged. Among the most characteristic elements of this law were the exclusionary rules. These were designed to prevent the jury from being misled by testimony and to keep the attention of the jury upon the issues of the pleadings. Again, these rules developed slowly; until the 1700's there was little distinction between ordinary and expert testimony. Gradually the distinction began to solidify.

The Opinion Rule

Among the most important exclusionary rules of the law of evidence is the "opinion" rule. In its modern form this rule states that a witness testifying on the issues before the court can offer only the "facts" personally and directly observed, not opinions, conclusions or inferences from those facts. Once the facts are known, a jury can be expected to draw the proper inference to form a correct opinion. A witness is permitted to express an opinion only when he or she has some special and relevant skill or experience that would aid the jury in arriving at conclusions based upon the facts. The purpose of the rule is to reduce or eliminate extraneous testimony before the jury.

This means that the testimony of witnesses on matters within the scope of common knowledge or within the shared experience of mankind must be generally confined to statements of concrete and observed fact, gathered by the use of an individual's own senses. A key witness can give an opinion about some limited matters, such as questions of identity of another, physical or mental condition of another, the dimensions of objects, the relative value of an object, or the identification of handwriting. The range of opinions is restricted and the "ordinary witness" is expected to testify in a manner supported by the facts.

While the opinion rule is simply stated it is not so simply applied; facts were not always distinguishable. Wide discretion in the testimony of witnesses is necessary when the dividing lines between fact and opinion are hazy and uncertain. A judge usually decides these matters by ruling upon lawyer's objections to testimony during the trial. A simple solution to problems of distinguishing fact from opinion is to let witnesses testify freely, leaving the application of the opinion rule to be developed further by cross-examination.

While the opinion rule limits the witness to matters actually within factual knowledge, the rule has an important exception: _it does not apply to expert witnesses_. An expert can offer analysis and opinions _beyond_ the mere presentation of facts. Expert testimony deals with matters beyond the knowledge and competence of the court. For this reason, the expert's testimony is accepted in ways that ordinary testimony is not.

The Expert With Firsthand Knowledge

Often experts are called as witnesses because of firsthand knowledge of the issues in court. The best examples are physicians who personally examine the claimants in personal injury cases. By direct observation and by direct application of medical diagnosis, a physician gains the factual knowledge needed about a litigant's disease, injury or suffering and can give an informed opinion in response to direct examination. But direct or firsthand knowledge is not limited only to medical matters. A hydrologist may have measured rainfall and runoff; his testimony could combine firsthand facts with professional opinions. Or a geologist might examine drilling core samples and testify about the geologic consequences of what he observed. Among the many other subjects of expertise resulting from direct or firsthand knowledge are questions about the authenticity of the age of writing and signatures; about the mental conditions or capacity of people; about the physical conditions of persons, including questions on personal injury and the cause of death; about the identification of persons or of things; about personal or real property values; about the cause of an accident or occurrence; about machinery or mechanics and construction or construction materials; about the identification or the explanation of photographs; about human emotions, affections, intentions or motives and other particular subjects (31 Am. Jur. 2d, Expert and Opinion Evidence).

Generally, the weight given to expert testimony is greater when a witness of accredited skill and experience forms an opinion from personal observation or examination of the subject at issue, and less when general conclusions are drawn on the basis of secondary or indirect study. The care and accuracy of an expert's analysis counts heavily, as does his independence and objectivity. The convincing or persuasive force of expert testimony may be weakened by a showing of strong personal interest or bias. Even when experts disagree, a direct experience or an analysis of the same questions in a different context usually makes strong testimony. Expert opinion receives as much weight as the circumstances can reasonably attach to it and direct experience is well regarded by both jurors and judges.

The Hypothetical Question

In addition an expert witness with no personal knowledge of the issue of a particular case can offer information to the court in the form of answers to questions. While a hypothetical question can be asked about a fact within the expert's direct knowledge, it is best used to gather expert opinion on facts beyond his direct knowledge. A hypothetical question presents claimed facts as hypotheses, then the opinion of the expert is asked. Since the hypothetical question is answered on a conditional statement of facts, a proper question cannot assume facts outside the evidence of the case. Nor can a proper hypothetical question call for the expert to supply both the premise and conclusion or to sum a wide range of testimony and offer an opinion on it. Designing clear, orderly and helpful hypothetical questions is a matter of great skill on the part of both examiners and cross-examiners. (Rosenthal, 1961).

A hypothetical question is an important tool for analyzing both factual situations and opinions on the implications or consequences of the facts. Usually courts control the form of hypothetical questions and regulate their content to assure that the question and answer helps the jury find an accurate and acceptable version of the truth. A properly framed hypothetical question will be stated so that the court will know what assumed facts led an expert to his opinion. The length of the question often depends upon the simplicity or the complexity of the issue. The question can have its source in the information contained in scientific books or monographs.

An expert's direct examination might contain hypothetical questions but the cross-examination by the opposing lawyer will almost surely contain them. An expert witness can be asked hypothetical questions for several purposes; among these are impeaching the credibility of an objective or recognized expert, testing the accuracy and reasonableness of expert testimony, contradicting prior testimony or demonstrating an opinion is or is not consistent with a theory relied on by the cross-examining party (31 Am. Jur., Expert and Opinion Evidence). Hypotheticals can be formed from theories that differ from the direct examination questions; as a general rule a cross-examining lawyer may frame a hypothetical question on any theory reasonably deduced from the evidence. An interrogator may even ask hypothetical questions based upon his own theories as long as the facts can be derived from the evidence submitted for examination. A cross-examining lawyer also may ask hypothetical questions either with facts from the direct examination or with facts included in the evidence but not part of earlier hypothetical questions. But no one can submit irrelevant evidence under the guise of cross-examination by a hypothetical question.

The hypothetical question can be an effective way to analyze the evidence. Skillfully used it can improve the quality of scientific, technical and expert information in the legal process. Yet the hypothetical question also can be abused. Questions can be vague, uncertain and ambiguous. Some hypotheticals may be misleading or speculative or an attempt to confuse the jury. Sometimes the question sets out the interrogator's conclusions, rather than the facts. The court has wide discretion in dealing with the form, content and use of hypothetical questions and the judge usually does not hesitate to use that authority. Hypothetical questions are subject to careful scrutiny in order to gain the positive benefits from use and to reduce the negative consequences from improper use.

Appropriate and concise hypothetical questions are difficult to develop and use, and trial lawyers appreciate experts who help prepare hypotheticals that elaborate facts and scientific theories. Experts who understand the form and purpose of hypothetical questions can help the lawyer ask better questions. Then the hypotheticals can be answered in a more orderly and consistent way.

Yet a question's substance is fully as important as its form, especially when the witness is a scientist or engineer with high levels of substantive knowledge. How do courts respond to the information in an expert's testimony? Can judges fully appreciate scientific or mathematical sophistication? Do they distinguish between valid scientific opinions that may differ? When experts disagree, how do judges decide which testimony to accept? The rules of evidence allow distinctions as to when testimony is or is not admissible. When an expert's testimony is admissible as evidence and how the decision is made should be a topic of concern to most expert witnesses.

Admissible Scientific Evidence

All valid evidence helps one find the truth, either in law or the sciences. Valid evidence has two major sources; the first is accurate and correct measurement. A procedure that measures what it should is more valid than one that does not. In addition, valid evidence is reliable; that is, consistent answers are obtained by repeated testing or experiment. Accurate, reliable evidence forms the framework for most scientific theories and laws, and the techniques for gathering valid evidence are important in the development of science (Walker, 1968). A science based on valid and reliable evidence has explanatory power, and is convincing to the analyst who develops it and the decision maker who uses it.

When courts use scientific evidence, legal rules are applied to determine whether or not the evidence is relevant. Relevant evidence is admissible in court; that is, the evidence can be accepted in the court's deliberation and used for its explanatory value. Scientific evidence passes two thresholds when used in court: first, the evidence must convince scientists of its validity, then it must convince the judge of its relevancy to the issue before the court. Evidence that survives these tests is admissible, it can be used in the determination of a legal decision.

The primary standard by which scientific procedures and techniques are determined to be admissible in court was first articulated by the U.S. Court of Appeals for the D.C. Circuit in the famous Frye v. United States decision (293 F. 1013 (D.C. Cir. 1923)). Deciding on the admissibility of polygraph evidence, the court held:

> Just when a scientific principle or discovery crossed the line between the experimental and demonstrable stages is difficult to define. Somewhere in this twilight zone the evidential force of the principle must be recognized, and while courts will go a long way in admitting expert testimony deduced from a well-recognized scientific principle or discovery, the thing from which the deduction is made must be sufficiently established to have gained general acceptance in the particular field in which it belongs (293 F. 1014).

The Frye decision sees an evolutionary process. Novel techniques pass through experimental phases under close scrutiny of the scientific community, then pass into a demonstrable stage where they receive judicial notice. A standard is set for distinguishing between the experimental and the demonstrable stages: the technique must be generally acceptable by the relevant scientific community. It is not enough for an expert, or several qualified experts, to testify that a particular technique is demonstrable. Frye imposes the specific burden of general acceptability (Giannelli, 1980).

The most important result of the Frye requirement is to shift the responsibility for applying the standard of admissible or non-admissible evidence. No longer must a judge make this determination unaided, no longer can a few expert witnesses determine what evidence is valid. Under Frye, the requirement of general acceptability in the scientific community assures that those most qualified to determine the general validity of scientific evidence will have the determinant voice. Scientists who know most about their procedures continue to experiment and to study with them. In doing so they become an informal but convincing technical jury.

If these scientists accept a procedure as valid for the problems they study, then a court can accept this determination in making findings of fact.

While the requirement seems straightforward, its application is often difficult.

If general acceptance is the standard against which the validity of scientific evidence is judged, then acceptance has two steps. First, the field in which the underlying scientific principle is found needs to be clearly delineated, and second, a determination must be made whether or not the principle is generally accepted by most of the members of that scientific community. Neither step is easy nor free from uncertainties.

The first problem is identifying an appropriate scientific discipline or field. In some cases the definition of a scientific principle is relatively easy for the origin of a principle is relatively clear and unambiguous. For example, the Uniformitarian Principle in geology states that rocks formed long ago at the earth's surface may be understood and explained in accordance with processes now in operation. In other words, past processes may be understood by studying present processes (Cox and Cox, 1974). This principle is closely associated with the field of sedimentary and physical geology. It is unlikely that another science, such as molecular biology, would offer a Uniformitarian Principle with similar meaning or application. If a lawsuit required an explanation of geologic processes, and if the Uniformitarian Principle offered valid evidence that could help assign liabilities, then the principle could be accepted by a court as a relevant explanation of physical fact and as admissible evidence. In this instance, the expert witness should be qualified to offer testimony about sedimentary processes; that is, he should be a geologist rather than a molecular biologist.

Many scientific fields are interdisciplinary and many scientific techniques are trans-disciplinary; consequently, selecting a field may be more difficult than it first seems. Hydrology is interdisciplinary and difficult to define as one science or another. As a discipline, hydrology includes concepts from geology, geophysics, and geochemistry. Hydrology includes civil engineering techniques from applied mathematics, computer science, civil engineering, agricultural chemistry, systems analysis, and soil physics. In addition, hydrologic analysis often incorporates the environmental sciences, economics, and sometimes water law. Defining any discipline is difficult, but defining an interdisciplinary field like hydrology is nearly impossible. Fortunately, hydrology has a "core" of techniques for measuring various sources and properties of water (Davis and DeWiest, 1966; Kazmann 1972). If a technique or the data generated by applying a technique are an important part of the evidence, agreement and acceptance of the technique can often be found among those who would be expected to be familiar with its use. These conditions have been accepted by previous courts, thereby eliminating the need constantly to define and redefine a discipline or a particular technique.

Once the relevant field or discipline has been identified, then the court must determine whether or not the underlying principle or technique has been "generally accepted" by members of that discipline. The acceptance need not be universal; it need only be widespread, prevalent, and extensive. More than a technique is at issue here, for widespread acceptance usually requires that the underlying theory be accepted as well (Boveridge, 1957). Establishing general acceptance is also easier if the theory is broadly recognized. Three methods of proving general acceptance are: expert testimony, scientific and legal writings, and judicial opinions.

So far courts are not unanimous on how to interpret the relationship between an expert witness and the theory or technique he supports. Obviously one expert does not speak for an entire discipline or scientific community. Courts generally look for corroboration. If two experts testify that a theory or technique is accepted, then courts are more willing to accept the testimony and to use it in an opinion. An expert's testimony also can be supported if the theory or technique is found in the literature. Courts are not validating the theory or technique, rather they are granting judicial notice to articles, texts, and reports, both in legal and scientific literature. This procedure also works well to establish a lack of general acceptance. Judges also often examine judicial opinions for guidance. If a theory or technique has been granted prior to judicial notice, it is more likely to be accepted by another court. This may seem to undercut the _Frye_ purpose--to grant those most qualified to determine the validity of a theory or technique the determinative voice--but judges have great discretion in accepting information form any source. Frequently that discretion is tempered by the experience and support of other judges facing similar questions (Giannelli, 1980).

Sometimes the scientific principles and procedures underlying a technique are so well established within the scientific community and are so uniform in application that the court will take _judicial notice_ of their existence. That is, it will not require the witness to testify as to their full theory and application before he testifies as to their results. For example, an expert fingerprint analyst, once his credentials as an analyst are established, may testify as to identification of the fingerprints he has analyzed without the necessity of fully explaining the scientific theory and techniques of fingerprint analyses. The theory and procedures have been proven to be so accurate and their application is so uniform that the court will simply recognize that they are the basis for the witness's testimony. The principles underlying the applications of radar, breath analysis, intoxication tests, firearms identification, and handwriting analysis have been considered judicially in a similar manner by the courts (Grannell, 1980).

So testimony can be admitted or excluded at trial by rules that have little to do with the substance of scientific theory or techniques, but nevertheless serve a legitimate purpose in a legal context. The _Frye_ rule illustrates that courts take a conservative approach to admitting scientific evidence: There is a preference for the exclusion of evidence of specious and unfounded scientific principles or conclusions at the expense of perhaps sound, but less accepted, novel scientific evidence. If alternatives exist which will allow admission of such evidence while addressing the judicial need for caution, it will be incumbent upon scientists to help the courts discover and adapt these procedures.

Special Problems of Quantitative Information

Courts are not alone with information problems: Hydrologists have them as well. Among the more important are those imposed by the mathematical nature of modern hydrology. Hydrology is a quantitative science: Although interdisciplinary, the science relies on empirical observation and mathematical analysis of physical processes. Hydrologists are educated to measure, to analyze, and to quantify. As a scientist, a hydrologist resembles the physical scientist and applied mathematician; that is, numbers are a stock-in-trade, units of scientific measurement are a daily discourse and equations are the tools with which understanding is sought. This may create problems for the hydrologist testifying as an expert witness because, although his expertise may be mathematical, mathematics may not be an effective way to present his evidence in court.

For several reasons, mathematical models should be used cautiously in courtroom testimony. The primary reason is so apparent that it is often overlooked completely: The courts are not scientific or technical institutions and judges, lawyers, or jury members are rarely mathematically trained. Scientific measurement and mathematical models are not the methods by which lawyers and judges seek explanation and understanding. Instead, the legal process is verbal. The law has strong written and oral traditions and legal logic has been verbal logic. Most of this development is unencumbered by mathematical expression. In order to participate comfortably and effectively in the legal process, an expert witness will speak the language of the discussion and not one more appropriate for other arenas or purposes. To speak higher mathematics in court without a compelling strategic purpose will probably gather to the expert suspicion, misunderstanding and distrust.

This is not to argue that mathematics has no place in court. Indeed one of the most interesting controversies presently discussed in the legal literature is whether or not mathematical methods have a place in a trial. Not all lawyers lack mathematical education; an increasing number have scientific or quantitative backgrounds. Recent arguments have detailed sophisticated analyses about the use of Bayesian statistics in evidence, with some arguing for the use of mathematical models (Finkelstein, 1978; Finkelstein, 1970) and others refuting the use of those methodologies (Tribe, 1971). Courts use mathematical and quantitative methods when these are appropriate to the trial. The determining factor is the courtroom need for such evidence. If a mathematical model would help the court understand complex issues, then an expert witness may be encouraged to provide a mathematical analysis. <u>This choice is by the court, not the expert</u>. So far, the weight of tradition in legal development is for verbal analysis in court, not quantitative.

When facing the court, an expert witness must make a presentation that is simple, clear and direct. The presentation must be effective the first time; a court is not a seminar of interested students or colleagues. There is little discussion or further explanation and elaboration of a point that is not clear when it is first said. This means that an influential and convincing expert is usually the one who strives to achieve trust and credibility with the court rather than precision and rigor. Trust and credibility are likely to increase when the use of mathematics or scientific jargon is decreased (Sabatier, 1978). The advocacy process guarantees the development of counter information and arguments by the opposing lawyers. The court will search for simple explanations to remember and easy rules to understand. If a presentation is relatively straightforward and clear, then the testimony will likely stick in the court's mind when the time comes to deliberate and choose.

Further problems remain with the use of mathematics in court. First, mathematical formulations of physical and behavioral phenomena tend to leave out unquantified or soft variables. This syndrome is a familiar one: if it cannot be counted, it does not exist or it is not important. After all, mathematical models are built specifically to focus upon the quantitative relationships among variables. Readily quantifiable factors are easier to process and hence more likely to be recognized as significant. The result, despite an appearance of accuracy and rigor, turns out to be a model significantly biased toward the quantified factors without an equal consideration of the non-quantified factors (Tribe, 1971).

Therefore, many mathematics models ask the wrong questions, or at least the wrong question from a court's perspective. Matters of identity and quantity--things that are objectively verifiable in the world outside the courtroom--lend themselves more readily to mathematical treatment than do matters such as intent and justice--issues important within the courtroom. In a laboratory questions of observable fact and quantity dominate; in a courtroom questions of right and wrong dominate. Value-laden judgments resist acceptable quantification, yet these are precisely the kinds of questions asked in the legal process. An important but unintended consequence of presenting mathematical models or quantitative proofs in court may be to shift focus away from the traditional concerns of the law, such as fairness, equity and intent, and toward such elements as identity, quantity, and occurrence. It is not clear that the marginal gains from having somewhat more precise answers would offset the losses from asking the wrong questions.

And mathematical models are seldom formulated or used to enhance understanding and participation by more scientists. Mathematical models are written for the expert who understands them because of education and professional capabilities. Part of the role of an expert is to understand what laymen do not understand. But mathematical sophistication is no substitute for an effective presentation by an expert witness. An effective expert must move beyond the formulas themselves and reveal the underlying truths so that the court can understand the full consequences of the decision it makes. Although mathematical models are valuable for scientific theory, their value so far in trials are marginal.

This is not an argument for the absolute avoidance of mathematics or mathematical models in court. At the extreme, such a position would be illogical for it would effectively invalidate much of the explanatory and predictive power of modern science and reduce the scientific or technical expert witness to the status of a storyteller. Instead, the scientist and engineer need to realize that the use of mathematics in court has great strategic significance, strategic in the legal sense rather than the scientific sense. Mathematics or mathematical models are used in court to gain legal ends. If a mathematical expression or model increases the chance that a lawyer will convince the court of his argument and win the lawsuit, then the expert will be encouraged to express and develop the mathematics.

Conversely, if a mathematical model would confuse rather than illuminate, or would somehow reduce rather than increase strategic advantage, then the expert will be encouraged to testify by speaking directly in ordinary language or layman's terms. The deciding factor is not the correctness or value of the equation or model, instead the decision to use mathematics is a matter of whether or not the result convinces the court of a litigant's claim.

Some experts are hired because of their sophistication in mathematical modeling, others are valued for their ability to explain scientific or technical ideas in straightforward language. Both are valuable for differnt reasons: both might be used in the same lawsuit to explain different aspects of the issue. Neither witness is inherently better than the other, nor is one more scientific or rational.

Each has legal value according to the trial strategy of the lawyer presenting his client's case. The witness who uses higher mathematics in testimony should realize the strategic implications of his choice, and the governing strategy is that of the legal process more than of the scientific process.

In summary, scientific information in the legal process will be used in ways that may surprise an expert witness and for purposes related to trial strategy rather than to scientific concerns. The trial is a different arena from that of a scientific institute or meeting, and information will be accepted or rejected depending upon its value to the legal process. The key to becoming an effective expert witness lies neither in mathematical sophistication nor in theoretical brilliance. Rather an effective witness understands the strategic role of scientific information, the nature of the court rules, and the need for convincing and persuasive evidence in the trial.

5 COURTS AND SCIENTIFIC INFORMATION

While the workings of the law mystify many scientists and engineers, the workings of science mystify many judges and lawyers. An envoy from science visits in court a strange and exotic culture. No calculated scientific method is used there; rather one finds an advocacy process with different fundamental rules. No data gathering. No hypothesis testing. No model building. In court an expert answers questions whose purpose is to explain some aspect of the lawsuit to the court. Explanations may require props: a blackboard or graphic display, a ball-and-stick molecule, a scale model of a process or industrial plant. Nevertheless, the method of examination and cross-examination is question and answer. Some questions are benign and direct, others are hostile and misleading. Some questions allow simple, straightforward answers because the facts and the relationships between facts are clear. Some questions are hypotheticals with more uncertain answers. For each question that tries to prove the contentions of the plaintiff, another will try to lead to the opposite conclusion.

Scientific expertise deals factually with almost every causal phenomenon except the ones most interesting to the law. A scientist abstracts to understand universal principles relatively free from human bias, interference or distortion. In contrast, the court seeks to resolve conflict with a decision that must recognize bias, greed and emotion. Scientists find truth in precision, belief and conviction. A scientist seeks to add an increment of firm knowledge to an evolving scientific discipline, a court seeks to settle each unique lawsuit within broad boundaries of fairness and justice. The differences are not cosmetic, each profession learns and practices a view of what is legitimate. The views rarely coincide.

This partly explains why the witness experience can be unpleasant. Isolated in the witness box, an expert has few, if any, books or notes and no friends in sight. To one side sits the judge, robed, remote, and aloof. If the trial has a jury, it sits on the other side, silent, impersonal, and quizzical. Both lawyers are relentless advocates, probing the premises, assumptions, and conclusions of the expert testimony. The opinion of the expert will not be accepted in the spirit offered, as knowledgeable, as objective, as true. Rather, the lawyers assume that all truths need careful examination of the premises leading to certain deductions. An expert will be challenged about academic qualifications, career patterns, and writings as well as analytic methods and conclusions. Surviving these challenges means that testimony is more likely to be acceptable to the court and useful in the determination of the legal outcome or decision.

Courts need experts to help resolve the complex issues of water and other environmental litigation. Expert analysis and testimony help the court assimilate information directly related to the case under consideration. Then, information based upon factual observation and analysis can form a foundation for the consideration of legal issues, such as how an injury might be redressed, or whether a discharge permit should be granted, of if a water right can be transferred for use in another location. Facts are an important part of the efficient and just resolution of these matters. If a court can be aided by the special or unique skills of an expert, then an expert witness is called. And as water uses and conflicts increase, more hydrologists will be needed to serve in this role, to help interpret the facts at issue in lawsuits, and to help develop sound decisions. Need motivates the increasing use of hydrologists and other scientists as expert witnesses.

But court tradition limits need. Although needed, scientific information must be structured for court use; that is, it must be appropriate to the issues being decided. Since a major role of the legal process is determining the responsibility for acts injurious to individuals or to society, scientific information furthering this determination will be readily assimilated. Information of this sort is extremely rare. Historically, science had little or no concern with whether or not a person was liable for the consequences of is or her activity; these value questions and moral choices were not science (Beveridge, 1957). Science was concerned with the definition of conditions under which a valid inference could be made to unobserved facts on the basis of evidence from observed matters of fact (Porro, 1979).

To some degree the tradition of an objective, neutral scientist is changing (Weinberg, 1978). Part of the change is internal; scientists are reexamining social responsibilities in a changing world. Most of the change seems to come from a need by society for scientific assessments of the consequences of alternative developments and the consequences of large scale technological innovations (Ravetz, 1978). Since modern water development involves significant technological skills and resources, these same consequences are a part of the analysis of most hydrologists. Social impacts, cost-benefit calculations and secondary effects require the application of scientific knowledge to problems once considered less important. While these problems are formulated and analyzed with scientific methods, the eventual solution is not straightforward. Rather, these problems are at a "trans-science" level: enough is known to state the problem in quantified or scientific terms but scientific knowledge is inadequate for a complete or acceptable solution (Weinberg, 1972). But even "trans-science" problems require a scientific analysis for understanding and the methods of scientific analysis are increasingly important to both public and private decisions. The institutional need for scientific advice and information is increasing rapidly (Coates, 1979). To provide scientific advice and information adequate for institutional demands is a relatively new role for many scientists, especially hydrologists. Once provided, the information becomes part of a complex problem of knowledge acquisition and utilization in the courts.

The problem of providing scientific information in the courts is both structural and procedural. As a structural matter, a scientist's analytic and scientific abilities are often inappropriately or ineffectively used. A court is an arena for advocacy and client-based partisanship. Courts resolve more questions of value than of fact with methods that rely more heavily upon judgment than analysis. The court seeks conflict resolution about judicial issues; science solves physical problems. The distinction is more than semantic; problems can be solved, issues cannot. A problem is solved by applying knowledge in a definite way, while an issue is resolved by striking a new balance among conflicting forces. Structurally the problem-solving approach of the scientist guarantees that purpose, however generous, and answers, however correct, will be mismatched with the purposes and solutions needed for a trial judgment.

Legal procedure also makes the hydrologist's information difficult to acquire and use. First, a trial is a two-party advocacy proceeding. Witnesses appear in court on behalf of either the plaintiff or the defendant; they appear neither as neutrals nor as representatives of professions, such as science or engineering. Witnesses' information supports the arguments of one side or the other in the trial. A good trial lawyer will not examine a witness without determining the value of the testimony to the case and will hesitate to allow a neutral or unsupportive witness to undergo direct or cross-examination by lawyers for the other side (Kraft, 1977). Testimony from witnesses provides most of the information about the conflict being tried in the lawsuit, and it is far too important for casual or careless use.

Experts are not accustomed to this rule. An expert's authority rests upon assumptions about scientific rationality: that scientific data is "objective", that scientific conclusions are based upon "rational" procedures and that scientific advice or testimony is a product of a superior evaluation scheme, the scientific method (Ezrahi, 1971). While science authorizes and certifies certain facts, the role of the scientist in court is a structured ambiguity. On the one hand, an expert speaks of general principles and, on the other, is often paid fees for testifying by one side or the other. Expertise is neutral but testimony is not. An expert witness is more than a mechanical conduit of technical data; rather an expert is a vital element in the legal process. An expert's testimony can help determine who wins or loses a lawsuit or who pays or is paid damages of how water law doctrines are understood and applied in the future. Realizing that scientific knowledge is useful to both sides for different purposes helps an expert appreciate how science is used in court (Nelkin, 1975).

In order to deal effectively with the problems of a lawsuit, scientific information must be specific to the problem, structured to be accessible and understandable and appropriate for a resolution of the conflict. It must be understood by the court. It must be scientifically accurate and strategically responsive. It must deal with the observable facts of a physical situation and with the human consequences of an authoritative decision; it must contain both quantitative analysis and qualitative judgments about what can and should be the answer. In other words, this information is more than a "solution," more than an application of a favorite model or theory and more than an analysis of the most easily understood part of the situation under review. This does not mean that an expert witness must carry a burden of information that is in excess of natural or technical abilities; far from it. But it is important that an expert understand the value of information in the context of the court trial to better serve as a witness and to provide contributions that result in better outcomes.

Legal decisions are seldom based solely on scientific or technical information. In addition to scientific information, a court is likely to consider: the statutory provision applicable to the issues, the legislative history of the statute, the legal precedents or past decisions in similar conflicts, the facts at issue in the case before the court and the methods of granting relief and treating fairly persons who are similarly situated. So information value in a lawsuit is more than a scientific or quantitative question, it is also a matter of statutory interpretation, legal precedent and fairness.

Experts usually provide information to the lawyer arguing the case of one of the litigants. The information has a specific strategic value in the lawsuit and will be used as such by the lawyer (Liebenser, 1962). But the expert witness can tactically assist in the lawsuit's overall strategy by understanding two important aspects of scientific and technical information. First, information is costly to acquire and use. The costs of gathering information -- time, money and skilled manpower -- are specific and immediate to the primary user, whether that user is the defendant, plaintiff or expert witness. But the benefits of that information are not so easily controlled. Once put in use, information diffuses rapidly. It can be modified and used by the other side in the trial, often to their advantage. It can add to the overall enlightenment or confusion of other members of the court proceeding, the plaintiff, the defendant, or the judge, or the jury. An expert's information also can be focused and controlled by the lawyer arguing the case. Great advantage is likely to accrue to the lawyer who marshalls expensive information with strategy in mind and disadvantage can easily fall upon the lawyer who is careless with this important resource.

The second fact is that all information is not equally important. Long, detailed accounts of technical matters are usually not an appropriate answer to direct or cross-examination. Yet incomplete answers also would be inadequate. Again, the amount and kind of scientific information that should be included in testimony is a tactical question to be determined by both the trial lawyer and the expert witness. While scientific information is tactically important, its strategic value accrues in combination with the other parts of the lawyer's presentation. In court scientific information must be direct and applied. Questions and answers must be directed toward the resolution of an issue.

Yet an expert witness has considerable discretion when testifying. How much to explain and how complete to make an explanation are problems that continually need addressing. Although all witnesses swear an oath to answer ". . . the truth, the whole truth and nothing but the truth . . .", things are not that simple. Some questions are straightforward with answers that are direct and clear; however, many important questions are considerably more complex. An expert may need to explain basic principles before a sophisticated answer can be given, to develop an explanatory background before a specific answer can be understood. An entire system may need explanation before a subsystem can be appreciated. An expert's testimony is both theoretical, in the sense of being based upon accepted scientific models and explanatory, in the sense of interpreting the issue under litigation with an application of scientific knowledge (Collister, 1968). It is not surprising that an expert's leeway in deciding what to explain and what to discount is of great interest to the trial lawyer. Indeed, the effective explanation of certain scientific points may be the crucial element in a trial strategy (Yellin, 1981).

Yet even if a scientist offers information or evidence in a clear and straightforward way, it may not be used or even accepted by the court. The judge in a trial has considerable discretion in deciding what information or testimony is or is not appropriate or pertinent to the matter under review. The information must be relevant; and even relevant information is not automatically admitted into evidence. Courts can exclude evidence if its value is outweighed by considerations of undue prejudice, if it misleads the jury, or if it unduly consumes time (McCormick, 1972). These considerations are beyond the expert's control. The expert's tasks are testimony and communication: the court will decide upon relevancy and procedural matters.

Once an expert witness realizes that the courtroom requires a different form of communication, the first task becomes clear: to develop arguments appropriate for the court by combining substantive expertise with trial strategy. In other words, an expert witness blends scientific fact with the situation at issue in the lawsuit, then explains or advises on the scientific consequences. Facts and themes are blended to become an argument. One argues to persuade. To argue is to construct facts and themes so that the conclusions are not only correct, compelling, and persuasive. Above all, a good argument is convincing (Meltsner, 1976).

A convincing argument is not neutral. Convincing arguments marshall evidence to support a conclusion (Baker, 1976). Many expert witnesses misunderstand the use of convincing argument. They believe that as a scientist their special duty is to offer testimony that is neutral, unbiased, and objective. But the accepted rules for neutral and objective scientific communication are not the same rules that communicate a convincing argument in the courtroom. While he must be accurate in his assessment, honest in his analysis, and objective in his reporting. The expert witness must also realize that his argument in the courtroom is not neutral, that it leads to a conclusion and that it is meant to persuade.

Some scientists may object to the idea of persuasive communication, fearing that persuasion is manipulation by emotion and prejudice. A scientist, they would say, should restrict testimony to only scientific matters and should state only facts. This is a mistaken view. It assumes that an expert can give acceptable and useful testimony by mouthing facts without considering strategy or the eventual trial outcome. All effective and skillful communication aims toward a single goal; that is, increasing the likelihood that the audience will embrace the message, understand it and act or decide differently because of it. Effective communication has an argument, only an argument persuades and convinces.

Does the expert witness need to accept an often unwanted role of advocate or partisan? Can the expert witness maintain credibility as a scientist while serving the court? How far need a scientist modify testimony in order to communicate it to the court? These questions arise every time an expert witness takes the stand. Each is answered anew whenever an expert witness testifies. What is needed is an appreciation that scientific facts and communication skills blend to become effective testimony. Scientific fact can be structured by rhetoric into enlightened testimony answering the needs of the court without sacrificing accuracy or validity. The first requirement is for scientists willing to take the time and effort to explain what they know in ways more acceptable to their audience. Once this effort begins, most experts find their credibility increasing, not decreasing, and they enjoy the satisfaction of keeping the courts make enlightened decisions using the best available scientific and technical information (McGaffey, 1979).

In addition to the difficulty of structuring testimony that is acceptable and effective in the court, the expert witness may face problems with his client. Sometimes an expert is faced with a client who knows what he needs to know. Under these circumstances, an expert witness has moral and ethical difficulties, for a hired expert presumably pipes the tune the payer wants played (McCracken, 1971). If the client or litigant has ideas or predispositions in conflict with a sound analysis or the professional recommendation of the expert, the expert witness is stuck in an ethical quandary. Should an expert testify to something he or she knows is less than the absolute truth?

Even this uncomfortable situation has alternatives which may be arranged along a continuum. First, the expert can ignore his or her client's dispositions and present the facts, expecting that the scientific truth will speak for itself. The interests of the client are held secondary to the unrestricted communication of scientific material. Second the expert can recognize his client's interests while presenting an analysis that addresses the issues before the court. Usually this means that the expert recognizes the client's interests and modifies his testimony so that the argument is more supportive of the client's position. Third, the expert can recognize the conflict and modify the testimony to conform completely with the client's or litigant's predispositions. The expert then becomes a complete advocate for the client's point of view.

None of these alternatives are attractive, but some are less attractive than others. All raise issues of ethics for expert witnesses. The first and last alternatives may result in the court and the client not learning from an expert witness what they need to know for a reasonable decision. The other alternatives require an expert to adjust his or her behavior or presentation in order to communicate more effectively. Of these two alternatives, the second is more appropriate. It calls for the expert witness to modify the style and form of the communication but to leave the content and message intact. Form is changed but not substance (Meltsner, 1978).

This modification requires strategic and tactical effort; indeed, testimony in this form is somewhat manipulative. No doubt some scientists will be uncomfortable with these ideas and will claim that changing form changes substance regardless of any tactical gain. But the point remains: without an effort to communicate, including the manipulation of form and style, an expert's testimony will be effective only in situations in which the analytic abilities, models and beliefs of the audience are shared and understood.

The Expert as Educator

The most important role for an expert witness in court is that of educator. Experts deal in a scarce commodity: knowledge. But knowledge is useful in court only after it has been offered and accepted, that is after the court has been educated. Knowledge based on fact is especially important in trials dealing with scientific and technical issues. Sharing this knowledge so that it might be used to reach rational and justifiable solutions is an expert's unique role. To be an expert is to be fundamentally an educator.

Education is more than collecting or presenting facts, experts educate the court by interpreting facts. Experts explain the significance of information and help to mold and direct the court's understanding of the problems at issue. Often, the expert finds that his role extends more broadly than first realized, for he or she educates not only the judge or jury but also the lawyer litigating the case. And an expert often learns by the experience.

In a strategic sense, the most important person an expert witness educates is the trial lawyer, for the lawyer is chief strategist of the lawsuit at trial. Only the lawyer has the required professional training in evidence, procedure, and legal substance to conduct the trial, from preliminary deposition to appeals. But in modern complex litigation, lawyers increasingly depend upon experts and professionals from other disciplines. Most lawyers are not technically trained, and most disputes over scientific or technical issues require the aid of specialists for understanding. The trial lawyer has little time to develop or expand any personal expertise because the demands of modern litigation are usually too pressing to permit such a luxury. In a few weeks of trial preparation the lawyer could not possibly become as expert as his witnesses in a technical discipline. But the lawyer needs to become expert enough to use the substantive material for certain purposes, to develop a case using physical data and to avoid pitfalls and errors of fact or judgment when handling scientific ideas and analyses. In other words, the lawyer needs a clear, straightforward and intensive education in those basic scientific principles that support the case. Usually that education comes from the experts hired to help in the litigation.

Education of this sort is of extreme <u>tactical</u> importance, for a lawyer's overall trial strategy depends upon the intelligent and effective use of both knowledge and resources. The hydrologist can prepare the lawyer to use scientific knowledge, stressing relationships that enlighten and support the basic underlying argument at issue, while avoiding those principles that might tend to confuse or obfuscate. Curiously, some lawyers cling to the notion that only they can be professionally involved in the trial, and experts should merely answer questions asked of them. While this viewpoint may have been acceptable in the past, modern litigation demands much more from both lawyers and witnesses. A good expert witness is unlikely to answer questions like a robot. Instead, a good expert is a valuable tactical resource, capable of understanding the legal process and anxious to help develop useful information. An expert who educates a trial lawyer can be a significant factor in the success of a lawsuit.

Courtroom Conduct

Assume that you are to serve as an expert witness and appear in court. Even after spending hours explaining hydrology and developing tactics with the lawyer, the realities of testimony may still offer surprises. While nothing substitutes for experience, some hints to avoid traumatic experiences may be helpful for novice and experienced witnesses alike.

Mistakes in Communication

Some scientists do not easily communicate in terms likely to be understood by lawyers or judges, while others find straightforward language and explanation relatively easy. Credibility is difficult to define but some witnesses seem able to establish rapport with a judge or jury much more quickly than others (McGaffey, 1979). The style of testimony and means of communicating chosen by the expert can be major factors in establishing credibility.

A frequent mistake made in communicating with a court is to stress method instead of results: emphasizing what was done in an experiment or an analysis instead of what was learned by the experience. This mistake arises from the natural frustration of trying to keep an audience's attention in the face of testimony that may be beyond a common level of understanding. But, total understanding is not the point. The court will need to understand only enough to accept the analysis presented by the expert. Acceptance is partly a matter of scientific correctness, but it is also the product of the presentation of the data, the way the data is analyzed and the credibility of the expert witness (McGaffey, 1979).

Many scientists think that recital of a detailed methodology is necessary to establish credibility with the court. Yet credibility has many additional sources, including credentials, reputation and outward appearance. Young experts often fall into the trap of trying to explain new and sophisticated analytic methods, often without extensive experience with the method. Generally, simple methods that fit the data and answer the problem are preferred to obscure, specialized and complex methods that are difficult to explain. Self-imposed limits upon the detail of methodological explanation are likely to reap rewards in the form of greater acceptance and credibility.

An expert witness should also remember that he is involved in education and therefore certain pedagogical principles should not be ignored. In a communication to an audience, the important points should be repeated several times. Repetition is good learning theory, as is use of certain forms of language. Short and straightforward statements in ordinary language are more quickly accepted than jargon or technical terms. And the extensive use of qualifiers or "weasel words" such as "probably," "tends to," "perhaps," or "maybe" can be avoided without loss in meaning but with gains in communication. The use of vivid examples can cause an important point to be remembered and used in the deliberations and decisions of the court.

The Appearance and Demeanor of an Expert Witness

An expert witness has an important advantage over an ordinary witness: he is a professional. While experts are not legally required to have certain educational experiences, degrees, licenses or registrations, all experts share the same functional characteristic--they "know what they are talking about." Knowing what you are doing, expertise, is the single identifying component of all professionals, and expertise and professionalism are nearly equal in the eyes of the public. Nothing can destroy that presumption more rapidly than an expert witness who steps to the stand without looking the part; that is, who gives an inappropriate image to the judge or jury either through dress conduct.

In the courtroom, as nowhere else, appearances count. Here is not the place for sandals, shorts and open-necked shirts, regardless of whether or not the air-conditioner is broken. Trim hair, neat clothes and clean fingernails make a definite impression. These clues indicate a person who cares what others see. Without the distractions of glaring inconsistencies between role and appearance, the substantive presentations of fact and scientific judgment can be much more effective.

Demeanor is more than clothes, it is an entire outward behavior. The soundest expert testimony can be rendered valueless by a witness whose demeanor is less than professional. The opposing lawyer will try hard to destroy a witness's composure knowing full well that this reduces the expert's credibility. An expert witness who dresses neatly, speaks clearly and calmly and sits straight in the chair has a psychological edge that allows an effective and impressive testimony. The expert must remember not to grin at the spectators, scowl at the audience or look at the opposing lawyer with obvious scorn. He should not divide his attention among the spectators, judge and jury but should devote it fully to the lawyer. Body language can plague an unprepared witness. Handwaving indicates nervousness. Avoiding eye contact makes an observer suspicious. And to say you are concerned, while inadvertently shrugging your shoulders, gives a strong visual clue that you are not concerned. Trials are visual; one must be careful of appearances and demeanor.

The Aggressive Cross-Examiner

Among the most frightening aspects of expert testimony is the grilling one expects during cross-examination by the lawyer for the other side. Actually, most difficulty with cross-examination comes from being ill-prepared and inattentive. Both can be neatly solved by preparing realistically for cross-examination and by taking a close and careful interest in the proceedings. Nevertheless, it is part of the lawyer's job to discredit opposing witnesses. Belligerent or abusive lawyers are increasingly a vestige of the past, especially in modern water litigation with its emphasis upon complex and expensive property issues.

The best way to deal with any aggressive cross-examiner is to keep cool in the hot seat. By analogy, aggression can be met by applying the basic principle of judo, the oriental martial art. In judo, an assailant's rushing attack is defeated largely by a redirection of the forces of his assault. A calm and quick-witted expert witness can do much the same thing to the verbal onrush of a cross-examiner. Never let yourself be provoked into a counter-attack. Retain your dignity and your sense of self-worth at all times; this will show the cross-examiner as a bully, something he can ill afford. Be polite, but resilient (Horseley, 1972).

Several pitfalls can be avoided. Three situations are typical of times when the expert witness should beware. First, beware of the cross-examiner who turns away as if finished, and then quickly turns back with the "Just a moment--one more question" tactic. Having put the expert at ease, he now offers a difficult question. The witness with his defenses down may be hit hard. It is best not to relax mentally until you have left the witness chair. The second pitfall is a cross-examiner who starts a series of short, rapid-fire questions calling for repeated "yes" answers. A careless expert witness develops a rhythm and may be caught unaware by a question calling for a "no" answer. Trying to recover and give the correct answer leaves the jury with the definite impression of carelessness or confusion. And third, watch out for the cross-examiner who prefaces his questions with "It's a fact." It may not be a fact, and if it is not, say so at once.

An expert witness will be asked about professional qualifications. Both education and experience are open to examination and cross-examination. In both instances, an expert's qualifications should be stated fairly and clearly. Do not inflate or pad your experiences; the true nature of what you have done may be brought out under cross-examination and shown to be less than remarkable. On the other hand, do not have an attack of excessive modesty. For example, an expert may be an outstanding authority in his field and clearly qualified to render expert opinions, without being the foremost authority in the world. He should not feel compelled to point this out to the court. For purposes of this trial at this time, he is the reigning expert. When asked whether one is an expert, a professional should be prepared to concede that indeed he or she _is_ one (Ames, 1974). The courtroom is no place to pull nasty surprises like excessive modesty on your lawyer.

A final pitfall continues to trap the unwary. There are many kinds of expert witnesses. One is the expert who has developed a position on a certain issue and never wavered, another is the expert who has never formed an opinion at all upon the issue under litigation and the last is the expert who has formulated an opinion and then changed it. Under cross-examination, the first may be accused of being closed-minded about alternatives, the second may be charged with laxity for never having considered the problem in question, and the third may be blamed for vacillation, for having no fixed opinion at all. Since these charges can cast doubt upon an expert's opinion, pretrial tactical sessions with the lawyer are an important opportunity to prepare for this kind of cross-examination and to cope with its implications (Gelpe and Tarlock, 1974).

How to Calculate a Witness Fee

One of the most difficult duties of the expert witness is calculating a fair fee to charge a litigant. If everyone were the same, the task would be easy: the same fee would be charged for everyone. But not all litigants are the same. At one extreme, most conservation organizations are able to secure expert witnesses for free, or for expenses, because many experts share the purposes and goals of the conservation movement (Sive, 1970). At the other extreme, some experts will consult with litigants strictly on a business basis, with high fees and all expenses paid. This type of expert often is a well-known leader or intellectual luminary. He is busy, and expects to receive substantial compensation for courtroom appearances.

Somewhere inbetween is the typical expert witness. Most experts are competent in a discipline, with a reputation somehow public enough to grant them name recognition. This expert may be approached by the litigant and asked to testify purely as a business matter, much like the more famous expert. Or perhaps a professional or social contact gives the litigant an opportunity to invite an expert to render professional services. The litigant usually explains that he or she is involved in a lawsuit in which the court requires independent expert views. The litigant, by now sure that expert's views are fair and reasonable, may ask that he share these opinions with the court (Ames, 1977).

Regardless of how the initial contact is made, the questions now begin. Can an expert charge a witness fee? If so, how much? Who pays the fee? What difference does a previous relationship with the litigant make? How can the expert assure payment of the fee? Even experienced expert witnesses are not always certain about the answers to these questions, and for good reasons. The courts themselves are not consistent about expert witness fees and how they are administered.

Two points are basic. First, the courtroom is no place to haggle about a witness fee. A potential expert witness with any question about the fee should settle the matter with the lawyer before going to court. Second, some states have laws that compel an expert witness to testify for standard witness fee or for nothing at all. In general, these guidelines apply: if a hydrologist is appearing as a normal or fact witness, he is entitled only to the standard witness fee. But if the hydrologist is appearing as an expert and providing an analysis beyond mere factual description, then he can set a reasonable fee. If an expert's testimony is vital to the case and if the expert spends considerable time and effort reviewing data and analyzing problems before the court appearance, then the lawyer commonly offers adequate compensation. If the lawyer does not make the offer, the expert should raise the fee question.

How much should an expert witness charge? And in what form should fees be computed? The answer to the first question depends upon who you are and your experience. The second question is tactically important; some forms of fees are more awkward and indefensible than others. Some experts contribute their testimony in the public interest, others charge as much as they can. Scientific experts have been reported to charge between $50 and $150 an hour for testimony (Sive, 1970). Naturally, the more famous and respected the expert, the higher the charge to testify. Also, the less time available to spend involved in court cases, the higher the fee. In actual practice, a fee will probably be negotiated between the expert and the litigant's lawyer. The negotiated amount will usually be slightly more than the lawyer wants to pay and slightly less than the expert thinks his or her time is worth.

Some other considerations are: the skill and standing of the expert, the difficulty or novelty of the trial issues, the importance of hydrology to a resolution of the dispute, the usual or customary fees for expert services and the benefits resulting to the litigant. A senior hydrologist with many years of experience, a highly respected name and undisputed expertise should rate a higher fee as an expert witness. Often a more experienced hydrologist can explain a given problem in less time than a junior hydrologist and the hourly rate would reflect the difference. If the lawsuit has difficult or complex questions that require hydrologic analysis for resolution, then the fee should be higher. Usually the fees for professional services are more or less standard in a region and can be taken as the guiding norm against which charges are measured. The best rule is to apply simple logic and reasonable judgment to the determination of expert witness fees (Jackson, 1975).

Fees should be quoted in amounts per hour: this is the tactically sound form. Some experts still quote fees in lump sums or in daily amounts but this has drawbacks. If a lump sum or daily fee is low and the expert puts in more time and effort than anticipated, then he comes out on the short end. If the fee is high, it gives ammunition to a cross-examiner in court. A tactic to make a witness appear naive or greedy is to ask what fee is being paid for an expert's testimony. If the expert's fees are low, the cross-examiner can insinuate that the testimony is also relatively valueless; after all, one gets what one pays for. If the fees are high, it can be insinuated that, ". . . $500 is mighty good pay for someone who's been on the stand less than half an hour. No more questions - you're dismissed." Although the fee may be legitimate, it might have a souring effect on those who labor hard for several days to earn $500. An expert should not be reluctant to reveal his fee because reluctance gives the impression of hiding something; however, an expert also should not give the impression that he or she is an intellectual mercenary selling testimony. A better answer to the cross-examination might be: "I'm not being paid for my testimony. I'm being paid for my time, my professional knowledge, my services in studying the hydrological facts of this case and for my scientific opinion based upon the facts." One cautionary note: If an expert witness is figuring a fee on an hourly basis, a cross-examiner may ask for an estimate of the total fee to be charged in this case, hoping that it will add up to an impressive sum. A good answer is: "My total fee depends upon how long you keep me here."

Hourly fees are tactically preferred. Most people work for hourly wages and can relate positively to hourly fees. An hourly fee implies that an expert is only being paid for exactly the time worked, and not for partial days. Also, hourly fees allow an expert to charge accurately for work done. By charging by the hour, an expert can charge for all of the time and services; that is, for data gathering, research and analysis, report writing, attending pretrial conferences, giving testimony at deposition hearings, and appearing court. If travel and lodging is involved in case preparation or testimony, then the expert should arrange for the transportation costs, hotel room, meals and tips to be paid by the litigant. If the attorney or the litigant cancels appointments on short notice, then the expert should figure in his or her fee the value of the time lost.

How and to whom are fees billed? Usually the expert is paid by the lawyer who calls upon him to testify. But the lawyer has no obligations to pay fees unless he or she specifically agrees to do so. The best way to avoid misunderstanding about who is to pay how much is to agree in writing upon a fee before expert services are performed. This can be done in letters, and the lawyer's letter to the expert should look like this:

> On behalf of my client, John Doe, I request that you prepare to testify as an expert witness in the case of <u>Doe</u> vs. <u>Doe</u>, scheduled for trial in such-and-such court at such-and-such time.
>
> My client and I realize that you will have to study the hydrologic problems involved in the lawsuit, make reports, attend pretrial conferences or hearings, provide advice and arrive at your scientific opinions, as well as testify in court. I agree on behalf of my client that you will be paid for time spent on those preparations and services at the rate of $____ an hour and that you will be compensated for any expenses you may incur in providing these services.

> In the event that the case is settled out of court, I agree on behalf of my client that you will be compensated for preparations and services provided by you up to the point of settlement.

The last paragraph is extremely important, since many civil cases never go to trial. Unless the expert witness has agreed that he will be paid for preparations and services regardless of whether court testimony is necessary, the expert may wind up empty handed (Horsley, 1977).

All other matters should be taken up with the litigant's lawyer, including billing arrangements, retainers and advance payment of expenses. Sometimes a losing litigant may avoid payment but the expert witness should still press for the fee, either through the lawyer or through a collection agency, if necessary. As a last resort, an expert has a contract in the letter stating the terms of service and the promised fee. If the contract is broken, an expert should consult his lawyer. This is a rare situation; an expert seldom has fee trouble if he reaches a clear understanding in advance on what is to be paid, when it is to be paid and who is to pay it.

In summary, an expert witness has many things to consider in addition to the substance of scientific testimony. An expert is an important tactical member of the legal team and must clearly understand the responsibilities and complexities of that role. An expert's main role is education, for the lawyer hiring the expert and for the court as well. Effective education requires more than substantive skill or knowledge; it requires careful presentation. An expert witness must look and sound professional while communicating in a direct and understandable way. If these tasks were not difficult enough, the expert must also keep cool in the face of hostile cross-examination. But for these services, an expert witness is usually compensated by a fair and reasonable fee. On balance, being an expert witness is one of the most challenging and rewarding tasks facing modern hydrologists.

6 CONCLUSIONS

An expert witness has important duties. Experts interpret science so that it may translate into the law. They serve as sources of scientific data. They analyze scientific information. They educate lawyers about scientific arguments to support a particular case, and they educate judges and juries about the meaning of science beyond the case at hand. They explain complex ideas to audiences of lay men and women. They help to assure a court decision based upon rational scientific knowledge. And they learn. In few other public services or opportunities for advising can an expert learn so much in such a short time. An expert witness may learn that the courtroom is not an engaging or supportive environment. This witness may prefer to remain in the field or the laboratory. Others may learn that the courtroom is a microcosm of a larger society in conflict. This expert might be fascinated by the tug and pull of emotional, economic, and legal forces during the trial. The courtroom might seem vital, alive, and stimulating. For both these witnesses a few concluding remarks are in order to hold the chapters of this book in a balanced perspective.

Expert Witnesses and Learning

The expert's education and communication role has more importance than can be appreciated in a single lawsuit or trial. Education is not a single level or a one-way process, it occurs on many levels at once. A scientist or engineer who educates a court by serving as an expert witness will learn a great deal from the experience. Two levels of learning are apparent: learning as an individual and learning as part of an institutional process.

The Expert as a Learner

What can an individual hydrologist (or scientist or engineer) learn by serving as an expert witness in court? Three lessons are immediately available, although much more can be learned by a perceptive expert. The first lesson is the operation of courts in a modern, democratic society. This is a valuable accomplishment in itself. Courts settle disputes over water, and by doing that, help determine present and future property rights and the ultimate allocation of water. Unfortunately, most scientists and engineers have limited opportunities for formal study of the courts and must learn about these institutions from first-hand experience. An expert witness has this opportunity and if the expert goes to trial relatively prepared, learning is quicker and easier. This book is a basic primer about court structure and operation, the use of scientific data in lawsuits, and the nature of legal reasoning. Hopefully, it will make as an expert's experience more insightful and meaningful.

Second, the logic and the methods of the law are fundamentally different from scientific logic and methods. The law has different goals, values, and assumptions. Law is historic while science is analytic. Law is subjective while science is objective. Yet both methods are essential to acceptable resolutions of water conflicts. Where one is weak, the other is strong. As a sound and equitable water law emerges from the current mosaic of legislation, statutes and trials, the knowledge of hydrologists will have a special place. An effective law must be grounded in reality. The most common way of capturing the special knowledge of science is by the efforts of expert witnesses.

The third lesson is more sophisticated and difficult to learn. The scientist or engineer is trained to understand problems by viewing them through a rational model. "If this happens, then that follows," says the scientist. When dealing with social or human behavior, the temptation is strong to apply the same methods of analysis and understanding. "If prices increase, then people will buy less water," or "If supplies are scarce, people will conserve for future uses." Unfortunately these formulations are at best incomplete and based upon partial analysis. Often they are incorrect. People are quite rich in thought and action, and human behavior has not yielded to a totally rational analysis. A human conflict, the basis for a lawsuit, is often more complex than a rational model can explain. So far, reality overwhelms models.

Therefore, scientists need to exercise caution. He or she is analytic by training and examines the quantified part of a problem because that seems more rational, more right. The normal and necessary way to solve a water resource problem is to understand it. But applying the scientific method to a situation is only one way of analyzing and only one way of understanding. More importantly, it is only one way of deciding the essential questions of what can or should be done.

Some problems are not solved by rational analysis. For these problems, different methods of coping have been developed. Recently an enlightened understanding of these alternatives has been advanced by social scientists. These are called interactive problem solving (Lindbloom and Cohen, 1979). Interactive problem solving works for situations in which the scientific method is inappropriate. Usually these are complex situations when causes and effects are uncertain or confused, or perhaps rights and duties are in conflict. Or a solution requires the application of the emotional or societal values of fairness, justice, or equity. In other words, many situations that a scientist or engineer might see in a courtroom are the kind of problems more appropriately resolved by interactive methods.

In contrast to rational approaches, interactive problem solving does not require the formulation of clear goals, the generation of data, the development of hypotheses and experimental studies, the choice among alternatives or variables and the ultimate "decision" that solves a given problem, although it may include part or all of this (Meltsner, 1976; Rourke, 1976; Benveniste, 1972). Instead, the interactive method absorbs ill-defined and uncertain problems into an established and accepted procedure, a process with historic legitimacy, social acceptance and ceremonial appeal. Solutions to the particular problem are reached by the application of the procedure, regardless of the eventual outcome. The procedure is the solution. The legal process is mostly interactive problem solving. The roles played in the drama of the trial help process a lawsuit to a conclusion. But trials are not plays (Ball, 1975). A trial is authoritative instead of merely entertaining; a court decision is implemented by the force and authority of government. A lawsuit is usually not structured by a conclusion known in advance and manipulated by an author or playwright. Finally, a trial is most often a unique situation resolved by the application of widely accepted legal precedent and statutory principles. Although a judge has discretion in how these principles and precedents are applied, the mere application of the principles is enough to legitimize a solution (Carter, 1979).

Interactive problem solving by courts has important virtues that make the process widely acceptable. Litigants come to court for a fair and unbiased hearing of their grievances. The court accepts evidence, hears testimony presented by advocates for both sides of the dispute, reaches a decision by the deliberation of a judge or a fair and impartial jury, assigns a corrective or punitive resolution within the provisions of the law and provides for further appeals and rehearings, if necessary. The process is applied in roughly the same fashion to thousands of lawsuits yearly, over a universe of disputes and claims, in courtrooms throughout the nation. And it works. The decisions are widely accepted and considered legitimate. Naturally, the winners in trials accept the verdict and believe that the right side won. Losers usually accept the verdict also. Even if they disagree with the decision, the process is accepted. The trial seems fair, balanced and legal. The interactive process solves people's problems, as much by how it works as by how it analyzes (Schwartz, 1976).

This is not to claim that there are two fundamentally different problem solving methods, analytic and interactive, and that they are wholly distinct and separate. Interaction and analysis are always mixed together, the methods complement each other. Without the need for better analysis in interaction, the scientist or engineer would have no future or little value as an expert witness in court. But while the methods work in concert, sometimes we need to remind ourselves that analysis is only part of a larger social relationship (Lindbloom and Cohen, 1979). The scientific professional needs to balance analytic and theoretic understandings with a realization that interactive problem solving is important for settling human conflict. As an expert witness, he has a unique opportunity to observe this decision system in operation and to reaffirm an appreciation for the world of human interaction and decision.

The Expert and Institutional Learning

Courts are not only collections of individuals who learn important individual lessons; nor are they only structures of authority and legitimacy. They can be seen as instruments to achieve social purposes. They are systems of communication and control. They are cultures. They are theatres for the play of conflicts of interest. When social psychologists, sociologists, system analysts, anthropologists, and political scientists examine courts, each inquirer finds a unique and interesting view of the institution. Increasingly, many perceive courts as institutions which learn (Argyris and Schon, 1978).

Clearly organizational learning is not the same as individual learning, even when the individuals who learn are members of the organization. Often organizations know less than their members.

Sometimes the organization cannot seem to learn what every member knows. Nor is organizational learning solely the prerogative of the person in charge who learns for the organization. In complex organizations such as courts, the people in charge, the judges, succeed one another while the organization continues and learns or fails to learn in ways often independent of any particular judge.

Courts develop the law by applying rules and processes to actions that come before the bench. The process has a ritual, but it is not mechanical. Lawsuits generate information in various formal ways. Lawyers advocate one explanation of the situation over the explanation, offered by the opposing lawyer. And juries or judges have the opportunity to reflect upon the information and the proceeding (Davis, 1980; Polsky, 1961). In thinking about the lawsuit, they learn from it. Then the decision is handed down and an opinion is written. The learning process may continue during an appeal with different actors reflecting again to detect or correct errors is an excellent description of organizational learning. Learning does not start from scratch with each lawsuit since the court also has past experience as a guide. The past is written strongly in previous opinions dealing with similar problems and judges consider themselves bound by these precedents. But lawsuits deal with each different conflict as it arises, and conflicts most often arise over new applications of the law or the rules that have not been previously settled. Here is enough flexibility for the court to engage in organizational learning: to develop an image or mental map of the situation, to detect a match or mismatch of outcome confirming or denying past legal theories and rules, and to correct error with a decision that invents new ways to apply the law.

Good information is vital: Information develops images or mental maps; information offers detection or matches or mismatches of outcome to expectations; information offers past rules and applications as a guide; and information serves as the vehicle for error correction. Some of this information comes from the litigants, some from the body of the law, and some from the development of the lawsuit. In a lawsuit on water, environmental or scientific issues, information also comes from the expert witnesses.

The expert is an important part of the institutional learning process. When testifying, an expert brings appropriate knowledge from science to the issues under discussion and review. Often scientific knowledge has been developed in different institutional settings for purposes other than court trials. The data is collected from field studies and observations. The hypotheses and experiments that use data are designed to ask and answer fundamental questions about the workings of the physical system. The analysis of results are done with sophisticated theory incorporating quantitative methods and, frequently, computers. The results are presented to scientific meetings, written into journal articles and books and taught to generations of students who will continue to expand the body of scientific knowledge during their careers. The information is valuable to society as well. In court, it must be used for enlightening the legal system. The use can range from background explanations to specific application of known facts and physical processes to the particular conflict that led to the lawsuit. The lawyers arguing each side of the trial will sculpt and filter the scientific data in a strategic way, seeking to explain the underlying social conflict as an argument for the litigant's contention. The other side will seek to discredit that contention, offering an explanation of the same facts supportive of the other litigant's case. Throughout these arguments, the court has an opportunity to learn about physical reality. Most trial participants will be unaware of much that the scientist offers. Learning from expert testimony should enable the jury or other court officers to form a legal decision more in line with scientific knowledge. The expert is a direct teacher of what the court needs to know in its decision process.

No process is perfect, especially not this one. Some lawsuits might lend themselves to an expert's analysis better than others. Some judges and lawyers might be more receptive to the tactical worth of an expert in the development of an argument or in the decision settling the litigant's dispute. Some might not understand an expert at all; others might understand perfectly. And some experts might be more effective witnesses than others. These criticisms are valid, but not determinative. The expert's role has undergone evolution since courts began to use scientific information in their deliberations, and it will continue to evolve in the future. It is important for a scientist to understand an expert's role and to particiapte in the court's institutional evolution so that, on balance, better decisions are made (Michael, 1973; Argyris and Schon, 1978).

The fundamental need seems clear: Better scientific information for court use, presented in ways that encourage instead of discourage its adoption (Gibson, 1974). Scientific and technical information that ensnares itself on the self-imposed traps of the naive and legally unaware is essentially wasted. In contrast, information structured by courtroom facts of life, such as the rules of evidence, the two-party advocacy process and the tactical development of a winning trial strategy, is as vital as it is novel. Regardless of discipline, an expert with a realistic idea of a witness's role is a valuable person.

So the learning takes place both ways at once. Courts learn from experts, and experts learn how to be more effective in court. In addition, learning by the court system is more pervasive and general than at the individual lawsuit level. Courts accumulate learning through written decisions. In turn, the decisions become part of future decisions on similar matters. The common law evolves through its lawsuits, reaching higher levels of greater understanding about specific problems. Decisions and precedent are a controlling synthesis of legal and societal knowledge. They regulate human disputes that reoccur in similar form, if not in similar detail (Vickers, 1978). One goal of the law is to resolve with consistency the same kind of conflict. But rules for change are important also. As problems change in society, so too change the lawsuits that courts decide. Precedent offers guides to the past, but decisions need an integration of new principles and knowledge. A court's learning is both backward looking and future-responsible (Michael, 1973). Integrating expert testimony into a broader decision system is an important task for both jurists and expert witnesses.

No one knows how this is correctly done. Information use is evolutionary as are the decisions from that information. So far, much of the role of an expert has developed in a haphazard and informal way. With further use of expert witnesses and a better understanding of their role, the conjunction between law and science can become an even more exciting arena of mutual appreciation and more rational legal decisions. The result might be a legal system that learns physical reality while developing and shaping law. The law, enlightened by science, might become a more responsive learning system, while pursuing goals of equity, efficiency, and justice. This is not far beyond our present experience. More informed and effective expert witnesses might help the courts obtain their goals sooner and easier. It would be a major accomplishment.

REFERENCES

Abraham, H.J. 1975. The Judicial Process. 3rd Ed. New York: Oxford Univ. Press.

American Bar Association. 1974. Law and the Courts.

Ames, M.P. 1927. Preparation of the Expert Witness. Trial Magazine, (Aug): 20-28.

Argyis, C. and Donald A. 1972. Schon. Organizational Learning: A Theory of Action Perspective. Reading, Massachusetts: Addison-Wesley Pub. Co.

Ball, M.S. 1975. The Play's the Thing: An Unscientific Reflection on Courts Under the Rubric of Theatre, Stanford Law Review, 28(1) 81-115.

Baker, S. 1976. The Complete Stylist and Handbook. New York: Thomas Y. Crowell Company.

Benviniste, G. 1972. The Politics of Expertise. Berkeley, Calif.: The Glendessary Press.

Beveridge, W.I.B. 1957. The Art of Scientific Investigation. New York: Vintage Books.

Black's Law Dictionary. 1968. 4th Rev. Ed. St. Paul, Minnesota: West Publ. Co.

Bockrath, J.T. 1977. Environmental Law for Engineers, Scientists, and Managers. New York: McGraw-Hill.

Bradley, M.D. and J.K. DeCook 1978. Ground Water Occurrence and Utilization in the Arizona-Sonora Border Region. Natural Resources Journal, 18(1) 29-48.

Brody, D.E. 1978. The American Legal System: Concepts and Principles. Lexington, Mass. D.C. Heath and Company.

Carter, L.H. 1979. Reason in Law. Boston: Little, Brown and Co.

Cataldo, B.F., K.G. Kempin, Jr., J.M. Stockton, and C.M. Weber 1973. Introduction to the Law and the Legal Process. New York: John Wiley & Sons, Inc.

Clark, R.E. 1974. Arizona Ground Water Law: The Need for Legislation. Arizona Law Review, 16(4): 799-819.

Coates, J.F. 1979. What is a Public Policy Issue? An Advisory Essay. Interdisciplinary Science Reviews, 4(1) 27-44.

Collister, E.G., Jr. 1968. Expertise: The Expert and the Learned Treatise. Kansas Law Review, 17(2): 167-179.

Cowan, T.A. 1963. Decision Theory in Law, Science, and Technology. Science, 140(3571): 1065-1075.

Cummins, R.G. and J.W. McFarland. 1980. Reservoir Management and the Water Scarcity Issue in the Upper Colorado River Basin. Natural Resources Journal, 17(1): 91-96.

Davis, K.C. 1980. Facts in Lawmaking. Columbia Law Review, 80(1): 931-942.

Davis, S.N. and DeWiest, R.J.M. 1966. Hydrology. New York: John Wiley & Sons, Inc.

DuMars, C. and Helen Ingram. 1980. Congressional Quantification of Indian Reserved Water Rights: A Definitive Solution or a Mirage? Natural Resources Journal, 20(1): 17-43.

Edwards, H.T. and J.J. White. 1977. Problems, Readings and Materials on the Lawyer as a Negotiator. St. Paul, Minn.: West Publ.

Expert and Opinion Evidence. 1967. American Jurisprudence, 2nd Ed., Vol. 31. San Francisco: Bancroft-Whitney Publ. Co., Supplement Pocket-Part, June 1979.

Ezrahi, Y. 1971. The Political Resources of American Science. Science Studies, 1(1): 117-133.

Finkelstein, M.O. 1978. Quantitative Methods in Law. New York: The Free Press.

Finkelstein, M.O. and W.B. Fairley. 1970. A Bayesian Approach to Identification Evidence. Harvard Law Review, 83(3): 489-517.

Furnish, D.E. and J.R. Ladman. 1975. The Colorado River Salinity Agreement of 1973 and the Mexicali Valley. Natural Resources Journal, 15(1): 83-107.

Gelpe, M.R. and D.A. Tarlock. 1974. The Uses of Scientific Evidence in Environmental Decision-Making. Southern California Law Review, 48(1): 371-427.

Gibson, G.D. 1974. Law in the Coming Years. Washington and Lee Law Review, 31(3): 495-504.

Giannelli, P.C. 1980. The Admissibility of Novel Scientific Evidence: Frye v. United States, a Half-Century Later. Columbia Law Review, 80(4): 1197-1250.

Holland, W.E. 1975. Mixing Oil and Water: The Effect of Prevailing Water Law Doctrines on Oil Shale Development. Denver Law Journal, 52(3): 657-694.

Hollister, C.W. 1976. The Making of England: 55 B.C. to 1399. 3rd Ed. Lexington, Mass. D.C. Heath and Co.

Horsley, J.E. et al. 1972. Testifying in Court: The Advanced Course. Oradell, N.J.: Medical Economics Co.

Hundley, N. 1975. Water and the West: The Colorado River Compact and the Politics of Water in the American West. Berkeley and Los Angeles: Univ. of California Press.

Ingram, H. 1973. Information Channels and Environmental Decision-Making. Natural Resources Journal. 13(1): 150-169.

Jackson T.P. 1975. Presenting Expert Testimony. Law Notes, 11(1): 21-24.

Kazmann, R.G. 1972. Modern Hydrology. New York: Harper & Row, Publishers.

Kilburn, P.D. 1976. Environmental Implications of Oil-Shale Development. Environmental Conservation, 3(2): 101-115.

Kraft, M.D. 1977. Using Experts in Civil Cases. New York: Practising Law Institute.

Lamb, B.L. (ed.). Guidelines for Preparing Expert Testimony in Water Management Decisions Related to Instream Flow Issues. Fort Collins, Colorado: Cooperative Instream Flow Service Group, Instream Flow Information Paper: No. 1. FSW/OBS--77/19 (July 1977).

Liebenson, H.A. 1962. You, The Expert Witness. Mundelein, Ill. Callaghan and Co.

Lindblom, C.E. and D.K. Cohen. 1979. Usable Knowledge: Social Science and Social Problem Solving. New Haven and London: Yale Univ. Press.

Lindbloom, M.P. 1977. Compelling Experts to Testify: A Proposal. The University of Chicago Law Review, 44(4): 851-872.

McCormick, C.T. 1971. Cases and Materials on Evidence. St. Paul, Minnesota: West Publ. Co.

McCracken, D.D. 1971. Public Policy and the Expert: Ethical Problems of the Witness. New York: Council on Religion and International Affairs.

McGaffey, R. 1979. The Expert Witness and Source Credibility--The Communication Perspective. Am. J. Trial Advocacy, 2(1): 57-73.

McLauchlan, W.P. 1977. American Legal Processes. New York: John Wiley & Sons, Inc.

Mellinkoff, D. 1963. The Language of the Law. Boston and Toronto: Little Brown and Co.

Meltsner, A.J. 1978. Don't Slight Communication: Some Problems of Analytical Practice. Policy Analysis, 5(3): 421-446.

Meltsner, A.J. 1976. Policy Analysis in the Bureaucracy. Berkeley and Los Angeles: Univ. of California Press.

Mervin, S. 1973. Law and the Legal System: An Introduction. Boston and Toronto: Little, Brown and Co.

Michael, D.N. 1973. On Learning to Plan and Planning to Learn. San Francisco: Jassey Bass Publ.

Mitchell, J.K. 1978. The Expert Witness: A Geographer's Perspective on Environmental Litigation. The Geographical Review, 68(2): 209-214.

Nelkin, D. 1975. The Political Impact on Technical Expertise. Social Studies of Science, 5(1): 35-54.

Patch, R.A. 1978. Compelling Expert Testimony: A Proposed Satutory Reform. The Hastings Law Journal, 30(2): 209-226.

Polsky, S. 1961. Expert Testimony: Problems in Jurisprudence. Temple Law Quarterly, 34(4): 357-377.

Porro, A.A., Jr. 1979. Expert Witnesses: Crossroads of Law, Science and Technology. American Journal of Trial Advocacy, 2(2): 291-304.

Ravetz, J. 1978. Scientific Knowledge and Expert Advice in Debates About Large Scale Technological Innovations. Minerva, 16(2): 273-282.

Revelle, R. 1975. The Scientist and the Politician. Science, 187(Mcr): 1100-1104.

Rosen, L. 1977. The Anthropologist as Expert Witness. American Anthropologist, 79(5): 555-578.

Sabatier, P. 1978. The Acquisition and Utilization of Technical Information by Administrative Agencies. Administrative Science Quarterly, 23(2): 396-417.

Sandifer, D.V. 1975. Evidence Before International Tribunals. Charlottesville: Univ. Press of Virginia.

Schwartz, M.L. (ed.). 1976. Law and the American Future. Englewood Cliffs, New Jersey: Prentice-Hall, Inc.

Silberman, L.J. 1975. Masters and Magistrates Part I: The English Model. New York University Law Review, 50(3): 1070-1180. Masters and Magistrates Part II: The American Analogue. New York University Law Review, 50(4): 1297-1372.

Sive, D. 1970. Securing, Examining and Cross-Examining Expert Witnesses in Environmental Cases. Michigan Law Review, 68(4): 1175-1198.

Thomas, W.A. (ed.). 1974. Scientists in the Legal System. Ann Arbor, Michigan: Ann Arbor Science Publishers, Inc.

Thompson, G.P. 1974. The role of the courts. In Federal Environmental Law, ed. E.L. Dolgin and T.G.P. Guilbert, pp. 192-237. St. Paul, Minnesota: West Publ. Co.

Trelease, F.J. 1979. Cases and Materials on Water Law. 3d. Ed. St. Paul, Minnesota: West Publ. Co.

Tribe, L.H. 1971. Trial by Mathematics: Precision and Ritual in the Legal Process. Harvard Law Review, 84(6): 1329-1393.

Vickers, G. Sir. 1978. Making Institutions Work. New York: John Wiley & Sons.

Walker, M. 1963. The Nature of Scientific Thought. Englewood Cliffs, New Jersey: Prentice-Hall, Inc.

Weatherford, G. and G. Jacoby. 1975. Impact on Energy Development on the Law of the Colorado River, Natural Resources Journal, 15(1): 171-213.

Weinberg, A.M. 1978. The Obligations of Citizenship in the Republic of Science. Minerva, 16(1): 1-4.

Weinberg, A.M. 1972. Science and Trans-Science. Minerva, 10(2): 209-222.

White, G. 1961. The Use of Experts by International Tribunals. Syracuse, New York: Syracuse Univ.

Wigmore, J.H. 1913. Select Cases on the Law of Evidence. Boston: Little, Brown and Co.

Wilson, J.Q. 1980. American Government: Institutions and Policies. Lexington, Massachusetts: D.C. Heath and Co.

Yellin, J. 1981. High Technology and the Courts: Nuclear Power and the Need for Institutional Reform. Harvard Law Review, 94(3): 489-560.

Younger, I. 1977. On Technology and the Law of Evidence. University of Colorado Law Review, 49(1): 1-8.

Court Cases Cited

Arizona v. California, 373 U.S. 546 (1963).

Folkes v. Chadd, 3 Doug. 157, 99 Eng. Rep. 589 (K.B. 1783).

Frye v. United States, 293 F. 1013 (D.C. Cir. 1923).

Continued from back cover

The Water Resources Monograph Series

Vol. 3
OUTDOOR RECREATION AND WATER RESOURCES PLANNING
J. L. Knetsch (1974), 121 pages
—Land use planning and recreation values have become important issues as today's population, income and leisure time has increased. This monograph aids the professional in incorporating these values in analyses of economic efficiency on both a local and regional level.

Vol. 2
BENEFIT-COST ANALYSIS FOR WATER SYSTEM PLANNING
C. W. Howe (1971), 144 pages
—Provides a helpful framework for project design with an emphasis on the quantifiable pros and cons. An important guide during this period of tight government budgets and capital markets.

Vol. 1
SYNTHETIC STREAMFLOWS
M. B Fiering and B. B. Jackson (1971), 98 pages
—Includes a summary of current proposals for generating synthetic streamflows, step-by-step numerical calculations and serves as a guide for their implementation and application in a variety of hydrologic engineering problems.

To purchase volumes or for further information on the Water Resources Monograph Series

Write to: **American Geophysical Union**
2000 Florida Avenue, N.W.
...gton, D.C. 20009
...62-6903
...r call toll free 800-424-2488